BIOACTIVE MOLECULES IN PLANT FOODS: AN EXPLORATIVE

BIOACTIVE MOLECULES IN PLANT FOODS: AN EXPLORATIVE

Edited by

Dr. Esha Rami

Delve PUBLISHING

www.delvepublishing.com

Bioactive Molecules in Plant Foods: An Explorative

Dr. Esha Rami

Delve Publishing
224 Shoreacres Road
Burlington, ON L7L 2H2
Canada
www.delvepublishing.com

Email: orders@arclereducation.com

ISBN: 978-1-77469-880-8 (Hardcover)

ABOUT THE EDITOR

Dr. Esha Rami is presently working as an Assistant Professor, in the Department of Life science, Parul Institute of Applied Science, Parul University, India. She did her Post graduated from Ganpat University, Ph.D. in biotechnology from Hemchandracharya North Gujarat University, in the year 2015. She has authored a number of national and international publications in reputed journals.

TABLE OF CONTENTS

LIST OF FIGURES

ABSTRACT

Bioactive compounds are available in small amounts in food sources, primarily in vegetables, fruit products, and whole grains, and offer health benefits past the fundamental nutritional value. Whereas pharmaceuticals have been shown to have severe side effects, bioactive ingredients from plant-based functional food sources have been exhibited to be effective in the treatment of different conditions with fewer side effects. Indeed, past research has uncovered the critical effects of bioactive elements from plant-derived functional food varieties with respect to the treatment of different conditions. This volume centers around plant-based functional food sources that show a recuperating capability and examines the various bioactive components and their pharmacological impacts. More importantly, the bioactive components of plant-based utilitarian food varieties are partitioned into six classes, in particular flavonoids, phenolic acids, alkaloids, saponins, polysaccharides, and others. Likewise, how these bioactive components show a healing impact is summed up into various classes, like the prevention of disease and improved immunity function. Generally speaking, this contemporary volume analyzes the use of bioactive elements from plant-derived functional food varieties as natural solutions for the management of disease and health improvement.

PREFACE

Plant-based food sources are a significant aspect of the human diet globally and a key source of organic active compounds. Their accessibility in a fresh form across the year is restricted, and because of their irregularity, processing becomes a vital element for preserving the accessibility of these health-promotion parts for everyday consumption. Today, traditional plant-based food sources, particularly produced from local items, are gaining momentum among consumers and researchers because of the most recent patterns connected to the re-utilization of natural resources from the general environment. The knowledge about their compound synthesis, including the presence and amount of bioactives that is connected to their organic action is fundamental. More often, different handling techniques are used, adjusting the composition of plant-based products. This is associated with the change of their structure and physico-chemical elements that may essentially moderate the biological impacts of end products.. Because of the various potential interactions of individual bioactives among themselves, and between bioactives and the components of complex food networks, the assessment of their process-moderated roles is pivotal for the assessment of their effect on health and disease, especially in model frameworks and complex food varieties. Moreover, the impacts of the synthesis of these processed items on digestibility do not seem to be completely discovered.

In this volume, different topics are covered, for example, the effect of food networks, food processing, and human digestion on the constitution and function of the bioactive ingredients comprising of plant-based food varieties items including botanicals, processed food varieties, functional foods, food supplements, plant-byproducts, food-based powders, and restorative plants. Bioactive compounds are present in trace quantities in food and plants including, vegetables, fruits, nuts, and oils. Large numbers of these bioactive compounds are useful for human health and have been extensively researched before. Bioactive compounds (BACs) with antimicrobial properties have been used for a long period as antimicrobial agents (AMAs) as pure compounds, plant extracts, and BACs. Nevertheless, insolubility, fragrance or flavor, and vulnerability to different handling conditions make it challenging to apply them as alternative biopreservatives in food networks. Researchers prescribe encapsulation as a method for the protection of their functions, cover or mask undesirable characteristics, and increase their bioavailability. Micro/nanoencapsulation methods can enhance the antimicrobial proficiency of BACs in food structures notwithstanding their natures or food qualities. In this volume, different BACs with antimicrobial functions, wall materials/carriers, and assembly technologies are captured.

Bioactive compounds have remedial properties upon intake, they lessen pro-inflammatory conditions, oxidative stress, and metabolic conditions. Epidemiological research demonstrates that a high intake of food sources rich in bioactive constituents with antioxidant function, including phytochemicals, vitamins and phenolic compounds, like flavonoids and carotenoids, affects human wellness and could decrease the risk of various disorders, like cancer, heart condition, Alzheimer's, diabetes, and stroke among other conditions. Bioactive compounds can modulate metabolic processes and exhibit positive properties, for example, cancer prevention, hindrance of receptor activities, enzyme restriction or induction, and gene expression inhibition.

Chapter

1 INTRODUCTION TO BIOACTIVE MOLECULES

CONTENTS

1.1 INTRODUCTION

Bioactive compounds are gaining popularity in a variety of fields, including modern pharmacology, geomedicine, plant science, cosmetics, agrochemicals, the food industry, nanoscience, and so on. This is an extremely promising area that is currently being fully developed. As a result, more research projects are being carried out to diversify the sources of bioactive compounds and improve their synthesis or salvage pathways.

Despite extensive research in a variety of fields, the definition of bioactive compounds remains hazy and ambiguous. The term "bioactive" is defined in a very small number of references, and most often in a brief paragraph that does not leave the confines of the specific field of study, limiting its applicability to a broader audience (Zoete, 2013).

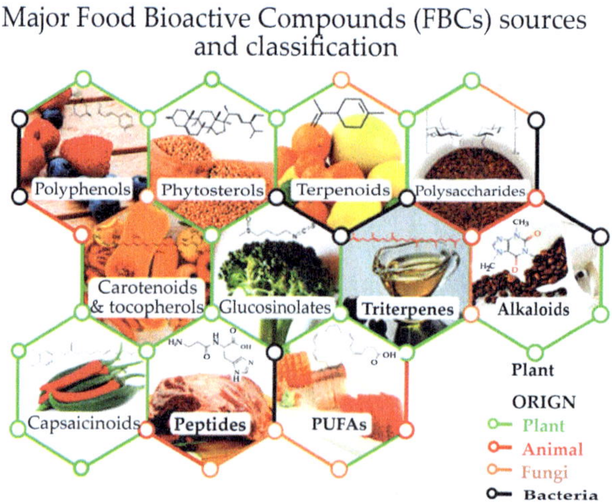

Figure 1.1. Schematic overview of structurally diverse bioactive compounds.

Source: Câmara et al., 2021. [accessed 7 Jun 2023]

The term "bioactive" is formed by combining the prefix "bio" and the suffix "-active." The prefix bio- refers to life and is derived from the Greek word "bios" (o-). Furthermore, the suffix "-active" denotes something dynamic, energetic, or involving an activity, all derived from the Latin "activus". This activity depicts any phenomenon that gives rise to a type of life, a working system, or a process.

In a strictly scientific sense, the term "bioactive" is a synonym for "biologically active." A bioactive compound is a substance that has biological activity. According to medical dictionaries, a bioactive substance affects, causes a reaction in, or activates a response in living tissue. If a substance or compound directly affects a living thing, it has biological activity. Indeed, depending on the substance, dose, or bioavailability, these compounds can have a wide range of effects, from beneficial health maintenance to healing to dangerous or fatal. The amount of bioactive substances consumed frequently determines whether an effect is positive or negative (Waldmann, 2013).

According to some authors, demonstrating these activities alone is insufficient to qualify a compound as "bioactive"; it must also have a positive impact on health. This definition distinguishes these substances from a large number of others that are bioactive but have side effects, are toxic, or cause cancer. This distinction has implications for the research experiences that should be planned to study bioactive substances.

In the first instance, "non-nutritional uses of nutrients" extend beyond the energy value of foods and their protein, fat, and carbohydrate contents to include other ingredients that can positively contribute to health preservation by altering the normal physiological function or increasing an organism's biological activity. There are bioactive food components, which are substances found in foods or dietary supplements that cause health changes but are not required to meet basic nutritional needs. Even if bioactive substances are present in foods or their components, it is critical to understand that they are not nutrients in this case. These definitions contradict those of Biesalski H-K et al.(2009), who provided the following definition in a "consensus" envelope: "Bioactive compounds" are natural substances that are both essential and optional to the food chain and have been shown to affect human health. Some definitions, on the other hand, distinguish between bioactive and essential compounds. According to the Encyclopedia of Food & Culture, the term "food bioactive component" refers to non-essential biomolecules found in foods that can modulate one or more metabolic processes, resulting in the promotion of better health. Similarly, "bioactive compounds" are extra, non-nutritional food components that are said to have positive health effects; essential nutrients are typically excluded. - Supplemental elements that are typically present in trace amounts in food. They have undergone extensive research to determine their impact on health. As previously stated, the term "bioactive food compound" or "bioactive dietary components" is typically associated with only beneficial effects on the organism. This definition considers the fact that bioactive foods affect health and, as a result, have biological value in addition to calories. This definition leaves out the potentially harmful effects (like toxicity, allergenicity, and mutagenicity), which are unquestionably the reflection of "bioactivity" in its broadest sense and have the potential (at least) to affect health positively (Waldmann, 2013).

In addition to this last statement, the use of neutral or ambiguous language (health effects, pretended, provided, putative) leaves a lot of room for the "negative" effects of various compounds on the living organism to be included within the definition of "bioactivity". Thus, "bioactive compounds" are naturally occurring, non-nutritive components of food plants that, when consumed, have the potential to improve or harm health. This method calls for greater use of non-food compounds. Several definitions appear to directly link the term "bioactive compound" to substances derived from foods (i.e., products found in food, substances that are part of the food chain), presenting it as a synonym for "bioactive food" and even more narrowly limiting it to plants or food crops. This is not the case, though! It's even possible to imagine "waste as a source of bioactive compounds" for substances derived from inedible sources. However, considering the criterion "origin" of these compounds, particularly in organisms that produce them, what interests us in this context are the fundamental ideas that underpin the definitions of related concepts. Bioactive compounds are made up of chemicals found in small amounts in plants and specific foods (such as fruits, vegetables, nuts, oils, and whole grains); these chemicals have actions in the body that can support health. Typically, nutrients are excluded from the definition of "plant bioactive compound" in plants. Typical bioactive plant compounds are secondary metabolites that are not required for a plant's daily survival (such as growth and develipment) but are essential for signaling, competition, defense, and other processes. Bioactive compounds have pharmacological or toxicological effects on humans and other animals (Zoete, V. 2013).

Thus, bioactive substances are found in organisms other than plants. These substances are also found in other living things and microorganisms, such as bacteria, mushrooms, and some animal species. The same is true for terrestrial (micro-)organisms as it is for marine (micro-)organisms. These also produce bioactive secondary metabolites, which may be useful substances. It should be noted that, despite advances in pharmaceutical chemistry and the appearance of new tools for chemical synthesis, which added another source of bioactive molecules, the ability to synthesize a diverse range of bioactive molecules did not emerge until the early twentieth century (Waldmann, H. 2013).

1.2 TOWARDS A UNIFORM PRELIMINARY DEFINITION OF BIOACTIVITY

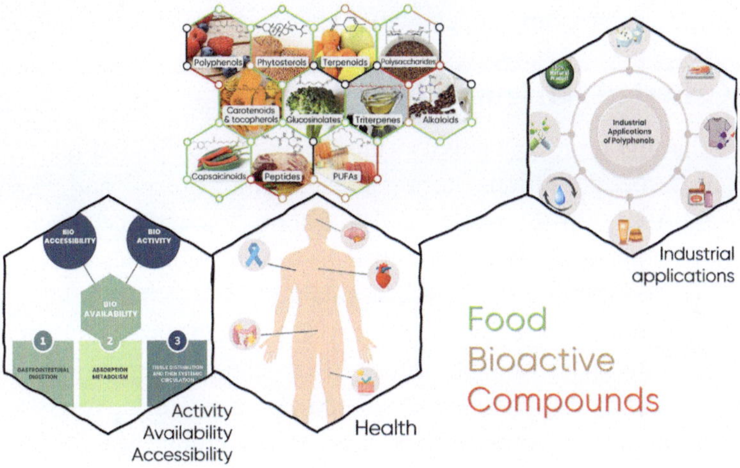

Figure 1.2. Representative examples of related bioactive molecules.

Source: Câmara JS, Albuquerque BR, Aguiar J, Corrêa RCG, Gonçalves JL, Granato D, Pereira JAM, Barros L, Ferreira ICFR. Food Bioactive Compounds and Emerging Techniques for Their Extraction: Polyphenols as a Case Study. *Foods*. 2021; 10(1):37. https://doi.org/10.3390/foods10010037 [accessed 7 Jun, 2023]

Right now, research on bioactivity potential is exploding, which may help to explain some of the "dynamism" that knows the definition of the term "bioactive," without ignoring the context of scientific studies that include different definitions. Identifying bioactive compounds and determining their health effects are active areas of scientific investigation. Despite ambiguous definitions and the frequent use of the term "bioactive," efforts to reach an agreement and identify some shared elements of the concept associated with the term "bioactive" have continued because there is no single definition for the term. According to Schrezenmeir J. et al. (2000), the definition of bioactivity typically has two limitations. The positive and negative health effects have already been discussed; the other requirement is that the bioactive component has measurable biological effects at physiologically realistic levels. Given the wide range of physiological functions that this bioactivity may influence, completing this mission will be difficult and will necessitate the participation of many specialties and disciplines at the same time. So, what exactly is a bioactive substance? All of the previously mentioned concepts must be combined to completely redesign a preliminary definition. This is accomplished by incorporating each concept into the puzzle one piece at a time (Zoete, 2013).

1.2.1. Criteria 1: Origin and Essence

Is it true that the term "bioactivity" implies that a compound is necessarily "bio"? In some definitions, terms like "biomolecules" and "occurs in nature" are used. On the other hand, the term "bioactivity" encompasses both synthetic and semi-synthetic substances. Many commercially available products (in cosmetics, medicine, agriculture, and other fields) rely on synthetic molecules as active ingredients. In most cases, the biological activity of bioactive compounds derived from natural sources and synthetic products is comparable. As a result, in addition to the discoveries made by isolating coumarins from hundreds of plant and animal species, synthetic derivatives have significantly increased the number of coumarin structures known to date. For many years, organic and medicinal chemists have been interested in the synthesis of coumarins and their derivatives due to their potential applications in the pharmaceutical, cosmetic, and agrochemical industries. Because of the undeniable importance of coumarin derivatives and their widespread use in recent years, numerous synthesis strategies and analytical techniques for coumarins have been developed, and numerous researchers have worked hard to create new compounds with simple substituents or more complex systems, such as heterocycles. A bioactive substance can thus be either natural or synthetic in origin (Waldmann, 2013).

The term "bioactive compound" does not refer to nutrients found in food or, more broadly, nutrients required by living organisms, such as primary metabolites. Other than the essential components of the definition of bioactive compounds, food (or sources of nutrition in general) is the subject of debate. Carbohydrates, water, proteins, fatty acids and lipids, vitamins and fibers, minerals, and oligo-elements are all food components. Bioactive compounds, according to some of the definitions given above, include soluble and insoluble fibers, numerous vitamins, fatty acids, flavonoids, and phytosterols. According to Liu (2013), these bioactive substances also include phytochemicals, vitamins, minerals, and fiber. According to some authors, such as Nahler (2013), vitamins and unsaturated fatty acids are not included in this classification because they are essential nutrients. To resolve this apparent contradiction, we use proteins, which have recently gained more recognition as physiologically active food ingredients and a source of bioactive peptides. Dairy products such as milk, cheese, and cheese made from cows are currently thought to be sources of bioactive proteins and peptides derived from food. In addition to providing nutrients, some food proteins, such as IGFs (insulin-like peptide hormones), lactoferrin, immunoglobulins, and β-lactoglobulin (beta-Lg) , have specific effects. Bioactive peptides can be encoded in amino acid sequence of a larger protein, and the beta-Lg is now a significant source of these peptides. Bioactive peptides are important in human health because of their antihypertensive properties, antioxidant and antimicrobial properties, and ability to lower body cholesterol levels. Following hydrolysis or fermentation, these peptides, which are typically composed of 3-20 amino acids, are liberated from the original protein (Zoete, 2013).

As a result, it appears that some nutritional components may have an extra-nutritional or non-nutritional function in the organism and exhibit bioactivity. Can such compounds, however, be classified as essential (nutritional role) once they exhibit biological activity? Because the same compound (or molecule) cannot serve two physiological functions in the same organism at the same time—nutritional (energetic metabolism and development) and bioactive (non-nutritional compounds are not essential in this case. This suggestion seeks to distinguish between two distinct processes: the first, which calls for the compound or molecule to be degraded to release the energy required for the organism's functioning and growth, and the second, which calls for the compound's (or molecule's) integral interaction with one or more components of the living tissue.

Simply put, a bioactive substance can interact with one or more components of living tissue. As a result, it circles back to itself, and this interaction (bioactivity) can occur regardless of the source of the bioactive compound, including food and non-food, and whether or not it is included in an essential nutrient. In this case, we can only define the bioactive compound by referring to its natural origin, but with the distinction suggested (Waldmann, 2013).

1.2.2. Criteria 2: Bioactive Effect Bioactivity

As previously stated, some definitions limit bioactivity to only having beneficial effects on the organism, particularly on human health. Examining various bioactive effects, including "negative" effects, for human benefit has always been and will continue to be a significant area of scientific study, not only in terms of health but also in a variety of other areas (agriculture, industry, and so on), with the ultimate goal of improving human life quality. The variety of effects that a bioactive substance can have on various biological organisms is determined by several factors such as chemical structure, dose, and so on.

An expected negative effect can be turned into a positive effect by changing one or more of these factors, or vice versa. Furthermore, a beneficial effect on one organism as a whole or on a specific tissue of that organism would not be beneficial to another, either directly or indirectly. The two most egregious examples that can be used here are (a) side effects (generally undesirable) that may appear in the same organism with the same bioactive compound, and (b) biocides, particularly those of natural origin, that would have a positive effect on a living organism (or tissue) by fighting against another that would be harmful to it.

As a result, depending on several variables that are typically related to three major changes, a bioactive substance may have very different effects on one or more components of living tissue: In this case, the living organism is the subject involved in the use of the bioactive compound, and the method of use (dosage, alone or in combination, etc.) is determined by the bioactive compound's physicochemical and biological properties (e.g., chemical structure, bioavailability, etc.).

1.2.3 Criteria 3: the quantity

Some definitions of bioactive compounds include the criterion "quantity" according to the definitions listed above: - Minor ingredients that may be harmful to human health; - Components found in trace amounts in foods; - Bioactive compounds, which contain chemicals found in trace amounts in some foods and plants. Giving the term "bioactive" meaning may be limited if the criterion "quantity" is included in the definitions (Zoete, 2013).

1.3 CLASSIFICATION OF BIOACTIVE MOLECULES

Bioactive molecules are chemical substances that interact with biological systems and have a significant impact on a variety of biological processes. These molecules are critical in fields such as pharmacology, biochemistry, and medicine. To examine their characteristics, functions, and potential applications, it is crucial to comprehend how bioactive molecules are categorized according to their chemical makeup and functional groups.

1.3.1 Classification Based on Chemical Nature

Bioactive molecules can be classified into several chemical classes based on their composition and structure. Major chemical classes include, among others:

- Alkaloids are a large class of naturally occurring organic compounds with alkaline properties and a nitrogen atom in the center. They are commonly found in plants and participate in a variety of biological processes. Alkaloids include for example, morphine, caffeine, nicotine, and quinine.
- Terpenoids: Terpenoids are a group of bioactive molecules derived from the isoprene unit. They are abundant in nature and have a variety of functions, including antimicrobial, antioxidant, and anti-inflammatory properties. Carotenoids, steroids, and essential oils are a few examples of terpenoids.
- Polyphenols: Polyphenols are phenolic compounds that contain multiple phenolic groups. They are widely distributed in plants and have drawn a lot of interest because of their antioxidant qualities and potential health advantages. Polyphenols, flavonoids, tannins, and lignans are a few examples (Waldmann, 2013).
- Glycosides are a type of sugar. Glycosides are compounds formed by the combination of a sugar molecule (glycone) with a non-sugar molecule (aglycone). They are widely distributed in nature and are essential to many biological processes. Some examples of glycosides are cardiac glycosides, such as digoxin, and cyanogenic glycosides found in certain fruits and seeds.
- Proteins and Peptides: Peptides and proteins are bioactive molecules composed of amino acids linked by peptide bonds. Among their many functions are enzymatic activity, hormone regulation, and structural support. Examples include insulin, collagen, and antibodies.

1.3.2 Classification Based on functional groups

Functional groups are atom clusters in a molecule that control its reactivity and chemical properties. Various functional groups influence the biological activities of bioactive molecules. The following are some common functional groups and the properties they are associated with:

- Hydroxyl (-OH) Group: Hydroxyl groups in bioactive molecules make them more likely to dissolve in water and form hydrogen bonds. Molecules with hydroxyl groups include alcohols, phenols, and sugars.

- Carbonyl (C=O) Group: Many bioactive molecules, such as aldehydes and ketones, contain carbonyl groups. These functional groups can participate in a variety of reactions, including oxidation-reduction and nucleophilic addition (Waldmann, H. 2013).
- Carboxyl (-COOH) Group: The presence of carboxyl groups in organic acids increases acidity and proton capacity. These structures are essential in biochemical reactions such as enzyme catalysis.
- Amino (-NH2) Group: Proteins, amines, and amino acids all contain amino groups. They can accept protons and act as bases, which influences how molecules behave biologically.
- Sulfhydryl (-SH) Group: Sulfhydryl groups are found in bioactive molecules such as thiols, where they aid in the formation of disulfide bonds. Disulfide bonds are essential for protein stability and folding.
- Ester (-COO-) Group: Ester groups are found in a wide range of bioactive molecules, including esterified fatty acids, ester prodrugs, and esterified vitamins. They aid in the stability and bioavailability of the compounds.

Bioactive molecules' biological activities and potential applications are influenced by their chemical structures and functional groups. Understanding the chemical nature and functional groups of bioactive molecules is critical for drug discovery, medicinal chemistry, and the development of therapeutic interventions. The classification in this article emphasizes the importance of the major chemical classes and functional groups found in bioactive molecules in the field of life sciences by providing a broad overview of those classes and groups (Zoete, 2013).

1.4 DIFFERENTIATING BETWEEN ENDOGENOUS AND EXOGENOUS BIOACTIVE MOLECULES

Bioactive molecules have a significant impact on human health and are required for a variety of biological processes. Understanding the origins and classification of these molecules is critical for understanding their effects and potential applications. The two main types of bioactive molecules are endogenous substances and exogenous substances. Exogenous bioactive molecules come from outside sources, whereas endogenous bioactive molecules are created by the body (Waldmann, 2013).

1.4.1 Endogenous bioactive molecules

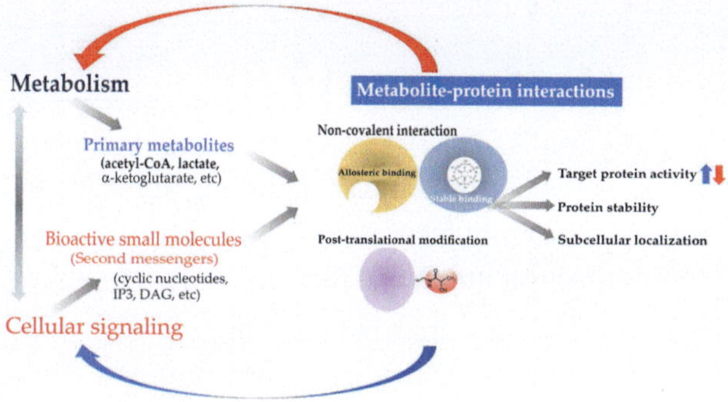

Figure 1.3. Biological actions of endogenous small molecules in the control of cellular signaling and metabolism.

Source: Kim S, Lim SW, Choi J. Drug discovery inspired by bioactive small molecules from nature. *Anim Cells Syst (Seoul)*. 2022;26(6):254-265. Published 2022 Dec 21. doi:10.1080/19768354.2022.2157480 Endogenous bioactive molecules are produced by the body and are involved in a variety of physiological processes. These molecules function as signaling molecules, immune response mediators, and regulators of cellular function. They are made in a variety of organs and tissues. Endogenous bioactive molecules that are commonly found in the body include hormones, neurotransmitters, cytokines, and growth factors.

Hormones are signaling molecules that control several bodily processes, including homeostasis, growth, metabolism, and reproduction. Endogenous hormones include insulin, glucagon, estrogen, testosterone, and cortisol.

Neurotransmitters are chemical messengers that are used to send signals between nerve cells. They are required for synaptic transmission, mood regulation, memory, and motor activity. Dopamine, serotonin, acetylcholine, and gamma-aminobutyric acid (GABA) are a few endogenous neurotransmitters (Waldmann, 2013).

Cytokines are small proteins that play a role in immune responses and cell signaling. They are in charge of cell differentiation, cell growth, and inflammation. Interleukins, interferons, tumor necrosis factors (TNF), and transforming growth factors (TGF) are a few examples of endogenous cytokines.

Growth factors are proteins that control cell proliferation, differentiation, and division. They play a role in embryonic development, wound healing, and tissue repair. Epidermal growth factor (EGF), platelet-derived growth factor (PDGF), and vascular endothelial growth factor (VEGF) are a few examples of endogenous growth factors (Zoete, 2013).

1.4.2 Exogenous bioactive molecules

Exogenous bioactive molecules come from outside the body and can be ingested through a variety of channels, including food, supplements, and medical procedures. These molecules are frequently used in therapy and have a significant impact on human health.

Many bioactive molecules are obtained through diet, most commonly from fruits, vegetables, herbs, and other natural sources. These substances provide numerous health benefits, including anti-inflammatory, antioxidant, and anticancer properties. Dietary bioactive molecules include polyphenols (such as resveratrol and quercetin), carotenoids (such as beta-carotene and lycopene), and flavonoids (such as catechins and anthocyanins).

Medicinal bioactive molecules can be provided by pharmaceutical drugs, herbal supplements, or other therapeutic interventions. These substances have undergone extensive research and are used for their specific therapeutic effects. Medicinal bioactive molecules include antibiotics (such as penicillin and tetracycline), analgesics (such as opioids and nonsteroidal anti-inflammatory drugs), and anticancer medications (such as paclitaxel and cisplatin).

Exogenous bioactive substances can also be derived from environmental pollutants or toxins. These molecules frequently have negative health effects on people and can cause several diseases or disorders. Heavy metals like lead and mercury, pesticides like glyphosate and organophosphates, and air pollutants like particulate matter and volatile organic compounds are a few examples of environmental bioactive molecules (Waldmann,. 2013).

1.4.3 Importance and consequences

Understanding their functions, sources, and implications for human health requires the ability to differentiate between endogenous and exogenous bioactive molecules. Endogenous molecules are required for homeostasis, regulation of bodily processes, and coordination of various bodily processes. They are an important component of the body's physiology. Disease diagnosis and treatment can benefit from an understanding of these systems and their dysregulation or deficiencies.

Exogenous bioactive molecules, on the other hand, provide opportunities for therapeutic interventions and dietary approaches to improve health outcomes. By identifying and harnessing the effects of exogenous compounds, researchers and healthcare professionals can develop targeted therapies and recommend dietary changes for a variety of conditions.

Finally, the origins, functions, and implications of endogenous and exogenous bioactive molecules differ. Exogenous molecules come from outside sources and can be used to treat patients, whereas endogenous molecules are produced by the body and regulate physiological processes. Understanding these distinctions is critical if we are to advance healthcare and improve human well-being.

1.5 EXTENSIVE STUDY INTO THE SYNTHESIS AND DESIGN OF BIOACTIVE MOLECULES

The creation and synthesis of bioactive molecules is an important field in drug research and development. Bioactive molecules are required for a variety of therapeutic purposes, including illness treatment, prevention, and improvement of human health.

The term "bioactive molecule" refers to substances with a specific biological activity, such as the ability to bind to specific targets within the body, control biochemical reactions, or have a therapeutic effect. These molecules can be derived from living things such as plants, animals, or microorganisms, or they can be synthesized chemically. Understanding the structure-activity relationship (SAR) and the molecular mechanisms underlying bioactivity allows scientists to manipulate and optimize molecules to produce desired therapeutic effects (Zoete, 2013).

1.5.1 Synthetic approaches to bioactive molecules

To synthesize bioactive molecules, target compounds are created by planning and carrying out chemical reactions. Medicinal chemists use a variety of synthetic techniques, including diversity-oriented synthesis, multi-step synthesis, and retrosynthetic analysis, to create libraries of compounds with potential bioactivity. Using retrosynthetic analysis, chemists can identify viable synthetic routes by breaking down a target molecule into simpler building blocks. Functional groups are sequentially modified through a series of chemical reactions in a multi-step synthesis to produce the desired structure. The primary goal of diversity-oriented synthesis is to create structurally diverse compound libraries that can lead to the discovery of novel bioactive molecules (Waldmann, 2013).

1.5.2 CADD stands for computer-aided drug design

The use of computer-aided drug design (CADD) has completely transformed the synthesis and design of bioactive molecules. Researchers can assess the pharmacokinetic properties of molecules and predict their binding affinities to their target proteins using computational techniques. They can also improve the activity of chemical structures. Methods such as molecular docking, molecular dynamics simulations, and virtual screening make it easier to identify potential lead compounds with strong therapeutic potential. CADD has significantly accelerated drug discovery while also reducing the time and money required for experimental synthesis and screening.

1.5.3 Enhancement of bioactive molecules

Certain aspects of bioactive molecules can be modified to improve their efficacy, selectivity, and safety. Medicinal chemists use combinatorial chemistry, high-throughput screening, and structure-based drug design to modify and improve molecules. Structure-based drug design can create bioactive compounds with higher binding affinities and specificities by using the structural details of the target proteins as a guide. Combinatorial chemistry allows for the rapid creation of compound libraries and the effective exploration of the chemical universe by combining different building blocks and functional groups. High-throughput screening makes it easier to find powerful bioactive molecules by comparing a large number of compounds to biological targets (Zoete, 2013).

1.5.4 Recent advances and prospects

Recent advancements in synthetic techniques, such as flow chemistry and green chemistry, have made bioactive molecule production more efficient and sustainable. Flow chemistry allows for continuous reactions, which have advantages such as higher yields, faster reaction times, and higher safety. Green chemistry emphasizes the use of sustainable reagents, the reduction of waste, and the use of environmentally friendly synthetic processes.

Furthermore, cutting-edge technologies such as artificial intelligence (AI) and machine learning (ML) have enormous potential for bioactive molecule synthesis and design. Large-scale data analysis with AI and ML algorithms can aid in the prediction of bioactivity, molecular properties, and synthesis pathways (Waldmann, 2013).

Figure 1.4. Future prospective on bioactive molecules.

Source: Kurek M, Benaida-Debbache N, Elez Garofulić I, Galić K, Avallone S, Voilley A, Waché Y. Antioxidants and Bioactive Compounds in Food: Critical Review of Issues and Prospects. *Antioxidants*. 2022; 11(4):742. https://doi.org/10.3390/antiox11040742

The synthesis and design of bioactive molecules is a critical component of drug development. Scientists work to create safe and effective therapeutics by understanding bioactivity, employing synthetic methods, employing computer-aided drug design, and optimizing molecules. Recent advances in synthetic methodologies and the incorporation of cutting-edge technologies provide future opportunities for the synthesis and design of bioactive molecules. The goal of ongoing research and development is to develop novel, potent bioactive compounds that will improve human health and well-being.

1.6 A BRIEF OVERVIEW OF THE APPLICATIONS OF BIOACTIVE COMPOUNDS

These naturally occurring bioactive substances are typically added to common foods or food products to provide wholesome calories and nutrients that are beneficial to the consumer's daily diet. These substances can coordinate key metabolic functions in the human body, such as free radical elimination, gene expression inhibition or induction, and receptor and enzymatic activity. Because of biological properties such as antimicrobial, anticarcinogenic, antimutagenic, antiallergenic, antioxidant, and anti-inflammatory actions, these compounds should be included in a healthy diet. They will help treat and prevent a variety of current and serious lifestyle diseases by being added to food. After carefully observing the lifestyle and needs of the populace in the present and near future, many commercial industries, including those in the pharmaceutical, food, and chemical sectors, are entering the business of extracting bioactive compounds.

The best sources of bioactive compounds are fruits and vegetables. They are biological compounds found in nature that may or may not be nutritional. They frequently aid in people's growth and development and reduce the risk of coronary diseases. Bioactive substances are now found in a wide range of industries, including food, medicine, and cosmetics. Plant extracts can be used as a functional component in the production of various food products, as well as in the cosmetics and pharmaceutical industries. Bioactive compounds are currently used by approximately 30% of the food and cosmetic industries, while 80% are used by the pharmaceutical industry. Bioactive substances are typically extracted directly from the raw source and added to it. They've also been extracted from marine microalgae and cyanobacteria. They are typically extracted using the solvent extraction method. The encapsulation technique is frequently used in the food industry to keep bioactive compounds functional. These substances are used in numerous industries. These compounds have mostly been used in the food, nutraceutical, and feed industries. A few of the compound's applications are discussed below (Zoete, 2013).

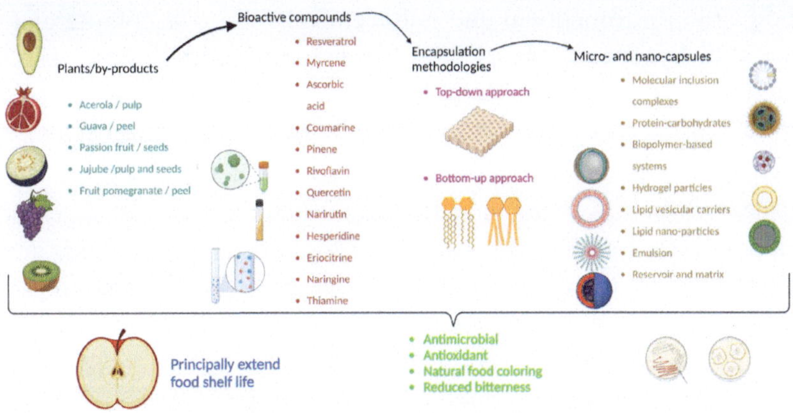

Figure 1.5. Various bioactive molecules along with their biological sources.

Source: Martins VFR, Pintado ME, Morais RMSC, Morais AMMB. Valorisation of Micro/Nanoencapsulated Bioactive Compounds from Plant Sources for Food Applications Towards Sustainability. *Foods*. 2023; 12(1):32. https://doi.org/10.3390/foods12010032

Antioxidant—Food processing and long-term storage conditions generate free radicals, which have an impact on food quality. This problem can be solved by scavenging free radicals and individual oxygen molecules, chelating metal ions (the problem-causing agent), stopping chain reactions (autoxidation), lowering oxygen concentration in the storage or processing system, and using other bioactive compounds that act as antioxidants (Waldmann, 2013). These substances will aid in the inhibition or induction of a specific type of reaction by regulating the activity of the enzyme.

Natural coloring agents, such as anthocyanins, carotenoids, curcumin, and squid ink, have the potential to replace all synthetic colorants, which are extremely toxic and non-nutritive to the general public while remaining highly stable in food systems. Carotenoids and flavonoids are primarily used to form pigments. In addition to these substances, other bioactive substances are used as natural dyes and food coloring agents. Anthocy

anin has long been used to treat conditions such as hypertension, diarrhea, and dysentery. It is also used to add color to foods and has been found to have therapeutic properties. These are primarily used in the food industry to provide desired food products with the necessary color. They not only have distinct color-imparting properties, but they also have a lot of potential for modification in functional and nutraceutical foods for commercial and domestic use. Lycopene, one of the main natural food colorings, is found in tomatoes, watermelon, and other fruits and vegetables. Because of its colorant properties, it is frequently used in human dietary products.

As flavoring agents in beverages, sweet foods, and chewing gum, natural flavors such as cinnamaldehyde and vanillin are used. Flavonoids are primarily found in beverages such as tea, wine, and beer. They are the primary sweetening ingredients in these beverages. They play an important role in the production of flavor and, to a lesser extent, sweetness in tea, wine, and beer. The two main types of flavonoids found in wine are anthocyanin, which gives a wine its distinctive color, and pro-anthocyanin, which primarily contributes to the astringent quality (Zoete, 2013).

According to antimicrobial agents, food is the best environment for the growth of millions of food spoilers and foodborne pathogens. Food spoilers attack the organoleptic qualities of food by metabolizing various food compounds to produce off-flavor, gases, slime, and other undesirable byproducts, whereas foodborne pathogens render the food unfit for consumption. Thanks to antimicrobial agents such as phenolic compounds, we can obtain food that is both healthy to eat and of a high calibre.

Agents of Texturization—Texturizing agents can include a variety of dietary fibers (soluble and insoluble), nutritive and non-nutritive gums, and other substances. Because of their remarkable ability to hold both water and oil, these compounds improve solubility and viscosity, which have a direct impact on the textural component of food products. Furthermore, dietary fibers will help you maintain a healthy gut and lose weight (Waldmann, 2013).

Agent of stimulation- Bioactive compounds, once extracted and purified, can be used to supplement functional and nutraceutical products, as well as a variety of staple foods (such as rice, salt, milk, and oils) and food products aimed at vulnerable populations. Agents that prevent DNA damage- Vegetables, plants, and cereals—all edible resources found in nature—contain a high concentration of bioactive compounds. Diets high in these compounds have been shown to reduce the risk of developing certain illnesses such as cancer, hypertension, cardiovascular disease, and many others. These compounds have properties that make them effective antimicrobial, anti-inflammatory, and anti-cancer agents. It also acts as a preventative measure against DNA deterioration. Citrus fruits, such as lemons and oranges, contain bioactive compounds that have been used to treat cancer. It also has chemo-sensitive properties. Antioxidant-rich natural resources are extremely effective at preventing DNA damage caused by radiation and other processes and reagents. Tough has not yet been used therapeutically because the compounds must be tested for toxicity before use.

Anti-inflammatory and anti-cancer agents- Flavonoids are members of the large phenolic compound group. It is split into six chemical groups: flavones, isoflavones, anthocyanin, flavonols, flavanols, and flavanones.. The most important group of these with anti-inflammatory properties is isoflavones. Isoflavones are abundant in soybeans and legumes. It not only has anti-inflammatory properties, but it also has anticancer, antimicrobial, and antioxidant properties. The most common isoflavones are genistein and daidzein. They typically act as phytoestrogens. According to the general mechanism, they bind to the mammalian estrogen receptor via pseudo-hormonal activities. It has anti-inflammatory and anti-cancerous properties because it binds to any powerful estrogen and blocks its pathways. It is critical in the fight against hormone-related cancers, such as breast and prostate cancer. Lycopene and other bioactive compounds are anti-inflammatory and anti-cancer.

Antioxidants- Phenolic substances, such as ascorbic acid and carotenoids, are the most common and abundant bioactive substances. Citrus fruits, Brussels sprouts, tomatoes, and other vegetables are the primary sources of ascorbic acid. Carotenoids are commonly found in foods such as papayas, sweet potatoes, and carrots. They are the most abundant source of antioxidants. The primary function of these substances is to protect the cellular membrane and its constituent parts from free radical damage. It can reduce microbial activity in some compounds. They can be used to preserve food in a variety of ways (Zoete, 2013).

1.7 A BRIEF OVERVIEW OF THE EXTRACTION METHODS

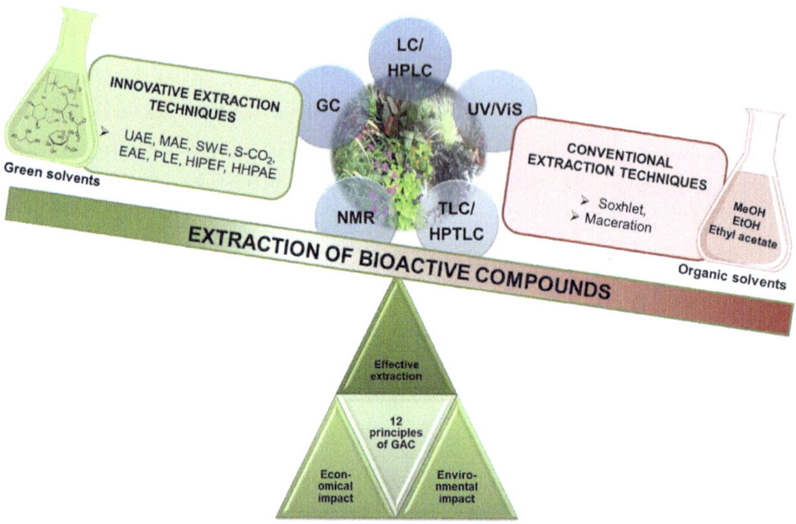

Figure 1.6. A typical process of bioactive compound extraction.

Source: Ivanović M, Islamčević Razboršek M, Kolar M. Innovative Extraction Techniques Using Deep Eutectic Solvents and Analytical Methods for the Isolation and Characterization of Natural Bioactive Compounds from Plant Material. *Plants*. 2020; 9(11):1428. https://doi.org/10.3390/plants9111428

Extraction involves mixing a pretreated raw material with an appropriate solvent to ensure that the bioactive compound of interest completely dissolves in the solvent and that other components, such as cells and tissues (other inert compounds), do not dissolve in the system (solvent + pretreated raw material). The extraction process is influenced by the type and quantity of raw material (plant, animal, microbes, etc.), as well as temperature, solvent type, pressure, and other variables. Each bioactive compound can be extracted using a different method. Conventional extraction methods (maceration, hydro distillation, and Soxhlet extraction) and nonconventional extraction methods (fermentation, controlled pressure drop, pressurized liquid, pulse electric field, microwave-assisted, solvent extraction, subcritical water, and supercritical fluid extraction) are broadly classified into two groups.

Maceration: During maceration, a suitable solvent is soaked in the ground material (which has been physically pretreated) in a closed vessel to extract the bioactive compound. The solution (which contains the bioactive compound and solvent) is separated while the remaining marc is pressed to extract the remaining solution (Zoete, 2013).

Soxhlet: The bioactive compounds are extracted completely using the Soxhlet method, which involves heating and condensing dried raw material and a specific solvent in a distillation unit repeatedly. It ensures that the desired compound is extracted into the solvent and collected in the distillation flask.

Hydrodistillation: To extract bioactive compounds from the target material, either boiling water or steam is packed inside of it. Water diffusion, hydrolysis, and heat-induced decomposition are all involved.

Pulse Electric Field: Exposing raw materials to a moderate electric field (between 500 and 1000 V/cm) for 10-20 seconds increases cell permeability. Exposure to a pulsed electric field ruptures the raw material's cell membranes, increasing the extraction process's efficiency (Waldmann, 2013).

Solvent Extraction: Polar and nonpolar bioactive compounds are extracted selectively using solvents such as methanol, acetone, ether, ethanol, and hexane.

Targeted material is placed in temperature-controlled, pressurized chambers with solvents such as methane, nitrogen, and ethylene to extract bioactive compounds. Supercritical states of solvents at high temperatures and pressures have better transportation properties due to their liquid-like density and gas-like viscosity.

Pressurized Liquid Extraction—This type of extraction is carried out at higher pressure (1450-2175 psi) and temperature (50-200 degrees C) to reduce the polarity of the solvents. Temperature can be used to match the polarity of the solvent to that of the desired compounds. Pressure will force the solvent into the raw material matrix.

Subcritical Water: This technique uses hot water (100-374 degrees Celsius) under high pressure (10 to 60 bar) to lower the dielectric constant of water to different values to replace organic solvents with water, which are typically used to extract bioactive compounds from the target raw material.

Fermentation is used to produce a wide range of bioactive compounds, either submerged (in a nutrient-rich liquid medium) or solid-state (on solid particles with or without direct water contact). Following the complete development of specific microorganisms on the raw material, these substances are produced as secondary metabolites. These compounds are produced when one of the essential nutrients in the media for their growth is depleted (Zoete, 2013).

Controlled Pressure Drop: The material is briefly exposed to saturated steam, followed by an immediate pressure drop that initiates volatile compound vaporization. The material expands as a result of additional instantaneous cooling caused by pressure drop, which prevents thermal degradation and protects heat-sensitive compounds. This increases mass transfer and recovery.

Microwave-Assisted—Microwaves help to evaporate moisture in cells by heating the water that is already present. These tissues experience pressure buildup during evaporation by rupturing their cell walls and organelles, increasing the porosity of the biological matrix. The increased porosity of the raw material increases the number of targeted compounds that come into contact with the solvent, ultimately increasing the yield of the same compound. Acoustic cavitation is a phenomenon that occurs when bubbles or cavities form in the liquid being extracted as a result of expanding cycles of ultrasonic waves. These bubbles begin to violently collapse after absorbing the energy from the ultrasonic waves, shattering the cells as they hit the surface of the raw material's solid matrix. These ruptures facilitate the release of bioactive substances (Waldmann, 2013).

1.8 BIOAVAILABILITY, BIOACCESSIBILITY, AND BIOACTIVITY

The three terms are related because a bioactive compound must be present in a living system to perform its primary function in that system. To be available in the living system, this compound must be released from the matrix it was ingested with. The bioaccessibility, bioavailability, and bioactivity of bioactive compounds in living systems determine their fate. As a result, detailed research on the compound's bioaccessibility, bioavailability, and bioactivity in the human body, as well as their respective food delivery systems, is required before their use in fortification and the production of functional foods.

The amount of a compound ingested in the gastrointestinal tract after it was released from the food matrix is referred to as bioaccessibility. The action of gastric secretion on the food matrix for the digestive transformation is required for bioactive compounds to be ready for absorption/assimilation into the bloodstream via intestinal epithelial cells.

Bioavailability refers to the number of bioaccessible substances absorbed by intestinal epithelial cells and transported to the circulatory system (via the blood or lymphatic circulation) to perform the desired physiological function. Not all bioaccessible bioactive compounds are absorbed because of the presence of other compounds or their interactions. Ascorbic acid and vitamin D, for example, increase calcium and iron absorption, whereas phytates decrease protein and mineral absorption (Zoete, 2013).

The ability of a substance to perform a specific physiological action when in contact with a specific type of tissue, such as an antioxidant, anti-inflammatory, or anti-diabetic, is referred to as bioactivity. The degree to which a compound interacts with other biomolecules during absorption and circulation, the impact of metabolism (gastric secretions), and other factors all influence its bioactivity at the targeted site. Nondigestible polysaccharides, oligosaccharides, and dietary fiber are examples of compounds whose bioactivity is independent of their digestibility and absorption.

1.8.1 Bioactivity and bioavailability influencing factors

Extraction Method: It has a significant impact on the bioactivity of the compounds. Temperature, pH, the type of solvent used, and the length of the extraction process can all have a significant impact on the functionality of these bioactive compounds. As a result, selecting the best method for each compound not only preserves its bioactivity but also maximizes its quality and quantity (Waldmann, 2013).

The Food Matrix: Bioactive substances frequently interact with other food matrix elements, which can increase or decrease absorption.

Chemical Formulation: The chemical structure of a target compound is one of the most important factors influencing its bioavailability. In this way, the structural modification of the bioactive compound affects how it is released, transformed, and absorbed during digestion, reducing its availability and/or preventing its release into the bloodstream.

Host-related factors- Diffusion is the primary mechanism by which bioactive compounds are absorbed from the intestine into the bloodstream and then into a specific organ or tissue. Because passive diffusion is ineffective for some of these substances, transmembrane transportation (active diffusion) is used to increase the permeability, uptake, and efflux of these substances. Because of competition for membrane binding sites with other substances, differences in binding site affinities, and other factors, targeted molecule absorption is frequently inhibited during active diffusion.

1.9 BIOAVAILABILITY, BIOACCESSIBILITY, AND BIOACTIVITY ENHANCEMENT TECHNIQUE

When bioactive compounds are added to food preparation and fortification, maintaining their bioaccessibility, bioavailability, and bioactivity until they reach the targeted site in the human body is a major challenge. Interactions with other compounds in the food matrix, as well as exposure to different pH, temperature, and oxygen conditions during processing and storage, harm the functionality and biological activities of these compounds. Some bioactive compounds are either not released from the food matrix or are ineffectively absorbed in the small intestine due to interactions or bonds with other molecules (Waldmann, 2013).

Encapsulation involves coating or trapping the liquid, solid, or gaseous bioactive compounds (core material) in a thin film (coating material) to stabilize their bioavailability and bioactivity. Out of the three different types of encapsulations (macro, micro, and nano), nano-emulsions, which are subsets of nanoencapsulation techniques, are the most effective methods for resolving the aforementioned issues. Nano-emulsions are dispersions of two immiscible liquids. Nanometric droplets of one liquid phase are dispersed throughout the continuous liquid phase of another liquid using this method. The entire dispersion is stabilized by an interfacial film of emulsifier or surfactant molecules between the two phases.

The continuous liquid phase (coating material) acts as a physical barrier in nano-emulsions. It protects against harsh processing and storage conditions, the stomach's acidic environment, digestive enzymes, and substances that bind to them and prevent their release. It also protects the bioactivity and bioavailability of other phases' nanometric droplets. According to numerous studies, the relationship between particle size and absorption through intestinal walls is inverse. Bioactive substances with a diameter of less than 500 nm are more easily absorbed and bioavailable. Nanosized particles of bioactive compounds created through nano-emulsions increase the surface area that favors passive transportation/rapid penetration through intestinal walls. Nano-emulsions have increased the bioavailability of omega-3 fatty acids, -carotene, tocopherols, curcumin, and essential oils while maintaining their bioactivity. For example, polylactic-co-glycolic acid nanoencapsulationincreased curcumin bioavailability by five to six times (European Food Safety Authority, 2011). Similarly, with smaller particle sizes after nanoencapsulation, fish oil bioavailability was improved (Zoete, \ 2013).

er

2 HEALTH BENEFITS OF DIETARY BIOACTIVE MOLECULES IN PLANTS

CONTENTS

2.1 INTRODUCTION

Plant-based diets are gaining popularity across the world due to their high-calorie density and nutritional value. The usage of these medications has grown in response to an increase in metabolic illnesses induced by a high-fat diet. Plant-based meals contain a variety of bioactive chemicals that may boost health in addition to nutritional advantages. Patients with cancer, cardiovascular disease, diabetic issues, gastrointestinal issues, immunological diseases, and neurological illnesses have been shown to benefit from carotenoids, polyphenols, phytosterols, biogenic amines, physiologically active proteins, and other phytoactive compounds. Previous research has shown that bioactive chemicals have antioxidative, anti-inflammatory, immunomodulatory, and intestinal barrier-improving capabilities, which all contribute to their capacity to attenuate the negative impacts of several human illnesses (Holmsen, 2010).

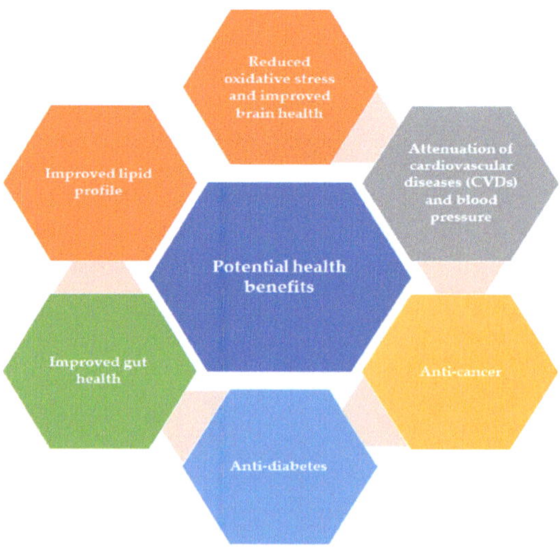

Figure 2.1. Potential health-promoting properties of bioactive components.

Source: Samtiya M, Aluko RE, Dhewa T, Moreno-Rojas JM. Potential Health Benefits of Plant Food-Derived Bioactive Components: An Overview. *Foods*. 2021; 10(4):839. https://doi.org/10.3390/foods10040839

Food provides vital nutrients as well as bioactive molecules or components that enhance health and illness prevention. A healthy lifestyle needs a high intake of plant-based meals, with at least 400 grams of fruits and vegetables consumed every day. Plants include a range of primary and secondary metabolites that impact human nutrition and health, in addition to the minerals necessary for a balanced diet. Although secondary metabolites are significant in both traditional and modern medicine, they are not necessary for plant development or function. Fruits and vegetables are high in micronutrients including magnesium, calcium, and potassium, as well as bioactive chemicals like polyphenols, dietary fiber, carotenoids, and vitamins.

There are roughly 5,000 distinct phytochemical substances in grains, vegetables, and fruits, and there are certainly many more. Whole grains, vegetables, and fruits have been linked to health benefits due to phytochemicals, which are bioactive, non-nutritional chemical components. Whole foods contain between 5,000 to 25,000 phytochemicals with bioactive characteristics, according to estimations. Fiber, calories, protein, protein calories, minerals, antioxidants, and protein calories are abundant in pulses and grains.

Vitamins, glycoalkaloids, and antioxidants are examples of botanical secondary metabolites. These compounds have been shown to have antioxidant, anti-inflammatory, cardioprotective, anti-obesity, and anti-diabetic activities. Fruit is a tasty plant component that can be sweet or sour. It is high in carbs, vitamins, and bioactive elements including fiber and phenols that may help prevent the development of some degenerative disorders (Holmsen, 2010).

Plant-derived polyphenols account for the great majority of naturally occurring antioxidants, including stilbenes, anthocyanins, flavonoids, lignans, and phenolic acids. Carotenoids and polyphenols both have anti-aging, anti-inflammatory, antiviral, antibacterial, and anticancer effects (Holmsen, 2010).

There has recently been a lot of interest in the bioactive properties of proteins or peptides obtained from food, which are controlled by amino acid sequences. Bioactive peptides are amino acid sequences that, regardless of nutritional content, have a favorable effect on a person's health and physiological performance. Bioactive peptides (BAPs) seen in natural proteins are thought to convert into inactive amino acid sequences. Nutritional preparation, fermentation, and enzyme-catalyzed proteolysis, on the other hand, result in fast BAP release *in vitro* or the human digestive system. Even though BAPs have been related to a decrease in the frequency of chronic illness in legumes, cereals have gotten less attention (Doğanlar, 2018)

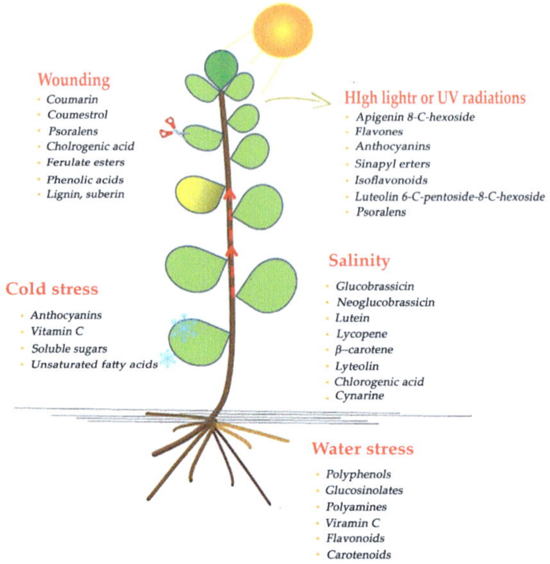

Figure 2.2: The bioactive molecule accumulation in plants.

Source: Toscano S, Trivellini A, Cocetta G, Bulgari R, Francini A, Romano D and Ferrante A (2019) Effect of Preharvest Abiotic Stresses on the Accumulation of Bioactive Compounds in Horticultural Produce. Front. Plant Sci. 10:1212. doi: 10.3389/fpls.2019.01212 [accessed 7 Jun, 2023]

According to a recent study, there is a lot of interest in bioactive chemicals taken from plants that may be helpful to human health. Plant-derived bioactive substances, such as physiologically active proteins, polyphenols, phytosterols, biogenic amines, carotenoids, and organosulfur compounds, have been shown in studies to cure several chronic illnesses. The scientific evidence supporting the use of plant-derived bioactive compounds in the management, prevention, or treatment of cancer, diabetes, cardiovascular diseases (CVDs), blood pressure, oxidative stress, mental health, gastrointestinal health, and lipid profile was investigated in that study. The identification of novel molecular targets for a wide range of human disorders has heightened interest in this subject. Few human intervention studies have been done to validate the alleged therapeutic advantages of plant-derived bioactive components. Extensive study, particularly clinical trials on people, is required for functional nutrients to have a major impact as a recognized pharmaceutical therapy.

Plants include a range of bioactive chemicals in addition to polyphenols, terpenes, limonoids, and carotenoids. Fruits, grains, vegetables, and millet are thought to give physiological advantages because of their high concentration of functional or bioactive components and nutritional value. Dietary supplements provide vital nutrients. This product is not designed to diagnose, treat, prevent, or cure any illness. These bioactive dietary components also include vitamins, fiber, minerals, fatty acids, and amino acids, in addition to polyphenolic chemicals. The vast bulk of the over 50,000 required components are multivitamins. Instead of being mixed with common foods or liquids (such as juices, stews, and so on), they are taken as a powder, tablet, or capsule. The principal phenolic chemicals responsible for the color of fruits and vegetables include leucoanthocyanidins, lycopene, carotenoids, betalains, and anthocyanins. They are referred to be "functional foods" because of their high antioxidant content and potential therapeutic advantages. The number of claims for fruits and vegetables is rising as a result of their importance in our meals and everyday life, dietary changes, and worldwide population expansion (Doğanlar, 2018).

Plant-based cancer therapies are preferred by patients since they are safer; nevertheless, not all naturally occurring substances are beneficial, thus this trend must be continuously watched. Cancer can be treated using *Taxus baccata*, *Taxus brevifolia* Nutt., *Podophyllum peltatum* L., and *Catharanthus roseus*. Hundreds of new cytotoxic components are discovered each year for use in the development of newer cancer-treating medicines. Understanding how a nutritious diet may improve our quality of life and assist in the treatment and prevention of noncommunicable illnesses is critical. Polyphenol-rich fruits, vegetables, and diets may help avoid metabolic syndromes and degenerative illnesses. Extensive study would be required to preserve bioactive components in naturally occurring plants. Extensive research and clinical trials are necessary before developing medications with low dosage effectiveness that can offer trustworthy information about the mechanism of action in the human body (Holmsen, 2010).

For millennia, herbal therapies have been employed. According to the World Health Organization (WHO), medicinal plants provide primary healthcare to nearly 80% of the world's population. Plant components with antibacterial qualities include roots, leaves, stems, and fruits, among others. Antibacterial characteristics are found in alkaloids, glycosides, tannins, and flavonoids.

Each of these secondary metabolites, both individually and together, suppresses microbial growth. The forests of Eastern Ghana include unique antibacterial plants from Western Africa. The Mampong Akuapem Plant

Medicine Research Center in Ghana collects, categorizes, and studies these plants. These medicinal plants include antibacterial chemicals that occur naturally.

Numerous RNA viruses, as well as bacteria, fungi, protozoa, and helminths, represent significant threats to world health. Viruses cause one-third of newly created and reemerging illnesses. The link between zoonosis and the transition from parasitic to viral illnesses between 1912 and 2013 highlights the importance of human socioeconomic activities as well as animal biodiversity in virus transmission. The three primary topics on which researchers are focusing to gain a better understanding of viral infection are the relationship between ecology, environmental change, and human societal dynamics and viral diseases, the possibility of person-to-person transmission of viral illness, and the treatment of viral disease. Because viruses may rapidly adapt to human hosts, developing precise therapies for viral illnesses is difficult. HIV is provided as an example (Doğanlar, 2018).

Pharmaceutical research has resulted in the development of a very effective HIV antiviral drug. HIV-positive patients may now treat their infection without acquiring acquired immunodeficiency syndrome (AIDS) if they follow their drugs as prescribed. Only 5% of patients can continue taking their drugs due to the high expense of therapy. Furthermore, little progress has been done in the hunt for antiviral drugs; so, therapeutic research is still needed to combat the prevalence of viral illnesses. The merging of medicinal chemistry and bioinformatics during the last decade has aided in the creation of drugs based on synthetic compounds. The two most visible challenges this technology is now experiencing are rising production costs and a decrease in the number of licensed medications. These problems are caused by insufficient distribution, metabolism, excretion, or toxicity.

Natural components, on the other hand, have been utilized as therapeutics for ages and are gaining favor as sources for innovative medications due to their accessibility and diversity. More than half of all chemical pharmacological entities were derived from or influenced by natural substances. Scientists are constantly studying the development of innovative medications using combinatorial chemistry and medicinal plants (Holmsen,2010).

2.2 PLANT-DERIVED BIOACTIVE COMPONENTS

Secondary metabolites, which are bioactive chemicals obtained from plants and produced metabolically, have exceptional medicinal effects, including antioxidative characteristics. The two most significant bioactive or phytochemical compounds that can benefit human health are phenols and carotenoids. Carotenoids are the most prevalent lipophilic chemicals in orange and yellow fruits and vegetables. These chemicals are ideal for usage in the food sector as pigments and health-promoting dietary supplements. Age-related macular degeneration, cataracts, cardiovascular disease, and sunburn-related diseases have all been proven to be delayed by zeaxanthin, lutein, and beta-cryptoxanthin. Furthermore, carotenoids have lately gotten a lot of attention because of their significant antioxidative action, which has been demonstrated to lower the risk of acquiring certain chronic illnesses.

Polyphenols, which are naturally occurring antioxidants, are found in vegetables, fruits, cereals, medicinal herbs, drinks, spices, and fungus, among other foods and pharmaceuticals. Polyphenols include anthocyanins, flavonoids, and polyphenolic acids. Natural antioxidants, such as carotenoids and polyphenols, have been shown to provide several biological advantages, including anti-inflammatory, anti-cancer, and anti-aging effects.

Figure 2.3. Examples of plant derived bioactive compounds.

Source: Nogueira GF, Oliveira RAd, Velasco JI, Fakhouri FM. Methods of Incorporating Plant-Derived Bioactive Compounds into Films Made with Agro-Based Polymers for Application as Food Packaging: A Brief Review. *Polymers*. 2020; 12(11):2518. https://doi.org/10.3390/polym12112518 [accessed 7 Jun, 2023]

Hydroxycinnamic acids (HCAs) are prevalent in apples, kiwis, cherries, plums, blueberries, and other fruits, ranging from 0.2 to 2 g/kg in fresh weight. Caffeine, a common phenolic acid, contributes approximately 75 to

100% of the HCAs observed in diverse fruits. Ferulic acid, on the other hand, is the most prevalent phenolic acid in cereal grains, accounting for more than 90% of the total phenol content in wheat grains. Anthocyanins are often employed as pigments due to their brilliant hues, but they can also be utilized to treat a range of human illnesses such as diabetes, cancer, cardiovascular disease, and inflammation. The health-promoting benefits of anthocyanins are mostly owing to their potent antioxidative capabilities. The tannins and flavonoids epicatechin, catechin, and gallocatechin are found in fruit pulp and soluble cell wall fractions. Salicylic acid, p-coumaric acid, ferulic acid, p-hydroxybenzoic acid, vanillic acid, gentisic acid, sinapic acid, and gallic acid are the main phenolic compounds present in bananas, whereas nitrogen-containing biogenic amines are generated via amino acid decarboxylation or aldehyde and ketone amination. Norepinephrine, serotonin, and dopamine have been found in banana flesh and rind. Vegetables and fruits are advised for intake due to their high concentrations of natural antioxidants such as vitamin E (tocopherols) and vitamin C (ascorbic acid).

Antioxidants regulate DNA damage, lipid peroxidation, and immune system activity, among other things. Plants have less dehydroascorbic acid than ascorbic acid. Polyphenols in edible and medicinal plants include phenolic acids, stilbenes, flavonoids, and lignans. Phenolic acids include benzoic acid derivatives such as hydroxybenzoic acids (HCA) and gallic acid, as well as cinnamic acid derivatives such as ferulic, p-coumaric, and caffeic acid. HCAs are found in larger amounts in food plants than hydroxybenzoic acids. Catechins are found in the highest concentrations in red wine and tea. Catechins are most commonly found as aglycones or gallic acid esters. Furthermore, celery and red pepper are good suppliers of luteolin and apigenin. Berries, plums, and eggplants exhibit violet, scarlet, or blue colors due to the presence of anthocyanins such as delphinidin, pelargonidin, and cyanidin. Daidzein and genistein are isoflavonoids found in legumes. Soybeans and soy-derived products are the most abundant sources of these chemicals in the diet. Quercetin is a flavonoid found in apples, scallions, and tea.. The flavanone naringenin, found predominantly in grapes and citrus fruits, has been shown to alter insulin transmission in the brain while preserving cognitive function. Proanthocyanidins, commonly known as condensed tannins, are the most significant flavonoid subgroup. They are polymers or oligomers of catechin (flavan-3-ol) containing interflavan linkages. They may be found in a variety of plant-based meals and drinks, including red wine, cereal, and fruits. Another type of plant-derived chemical is organosulfur compounds (OSCs), which are bioactive chemicals. Onions (Allium), which contain high levels of OSCs, have been linked to the protection of several other chronic illnesses, including diabetes, obesity, cardiovascular disease, and metabolic disorders. According to new research, isoalliin, the major sulfur molecule in onions, and other precursors such as S-methyl cysteine sulfoxide (methiin), are precursors to a wide range of flavor-enhancing and health-promoting chemicals (Holmsen, 2010).

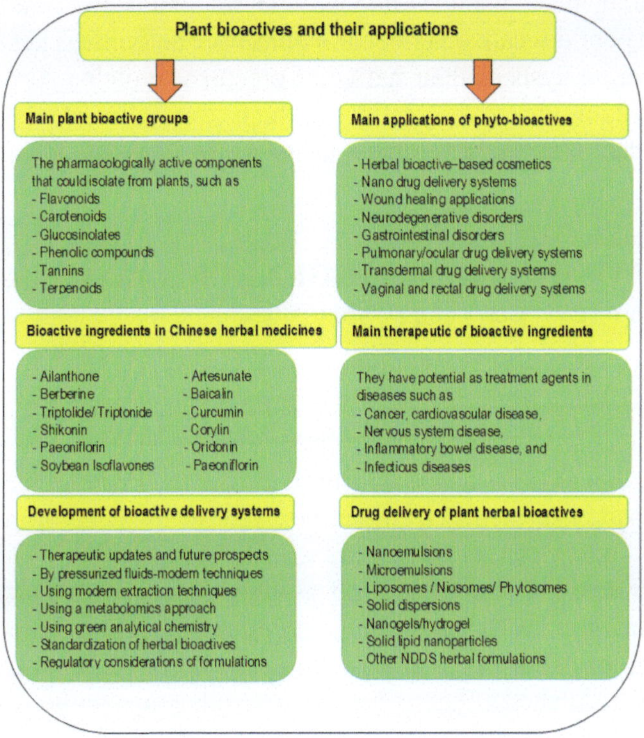

Figure 2.4. Medicinal plants have many bioactive compounds.

Source: A Comparative Photographic Review on Higher Plants and Macro-Fungi: A Soil Restoration for Sustainable Production of Food and Energy

Regular consumption of orange juice (OJ) reduces oxidative stress and inflammatory reactions, according to a previous study. It has been indicated that OJ flavanones, specifically naringenin-7-rutinoside and hesperetin-7-O-rutinoside, have a variety of health benefits, including enhanced lipid profiles. The Allium plant genus comprises the garlic plant genus, which has a lengthy history of use as a medicinal herb. Their anti-cancer and anti-diabetes properties have been investigated and established. According to scientific research, these health benefits are the result of bioactive chemicals such as phenolic compounds and organosulfur compounds. The most well-known component of the Mediterranean diet is virgin olive oil (VOO), which has gained popularity due to its nutritional and flavorful qualities. The majority of the health benefits of VOO are attributable to its phenolic compounds, which include hydroxytyrosol, tyrosyls, flavonoids, vanillic, and p-coumaric acids (Doğanlar, 2018).

In general, plant peptides and protein fragments are smaller than 10 kDa. These peptides can either occur naturally or be synthesized through the enzymatic proteolysis of their native or precursor proteins. Several dietary sources contain bioactive peptides, which can influence physiological processes that may be advantageous to human health when consumed (Doğanlar, 2018).

2.3 PROMISING HEALTH-BENEFICIAL ATTRIBUTES OF BIOACTIVE COMPONENTS

Due to their propensity to donate hydrogen atoms or electrons, phenolic compounds are well-known as important antioxidative or free radical inhibiting agents. Antioxidants in fruits and vegetables can reduce the risk of cardiovascular disease, arthritis, cancer, neurological diseases, atherosclerosis, and mental dysfunction. Antihypertensive properties have been identified in rice, barley, oat, wheat, and cereal-derived bioactive peptides. Protease digestion of food proteins results in the production of peptide sequences with lipid- and cholesterol-lowering properties. The consumption of Allium vegetables such as scallions, garlic, and onions has been linked to diabetes, heart disease, and other metabolic disorders. In addition, their consumption has been linked to a reduction in a variety of malignancies.

Biological properties of plant-derived bioactive compounds include anti-cancer and anti-diabetic activity, the amelioration of cardiovascular diseases, the maintenance of oxidoreductive balance and brain health, the improvement of blood lipid profile, and the regulation of gut health, as evidenced by the scientific literature (Holmsen, 2010).

2.3.1 Attenuation of cardiovascular diseases (CVDs) and blood pressure

The World Health Organization (WHO) estimates that cardiovascular diseases (CVDs) account for the death of 17,9 million persons annually.

This number is projected to nearly triple by 2030. CVDs are heart and blood vessel diseases, such as coronary cerebrovascular disorders, heart disorders, and rheumatic heart disease. The kallikrein-kinin (KKS) and renin-angiotensin (RA) pathways regulate sodium (Na) levels and blood pressure (BP). Angiotensin I is converted to angiotensin II by angiotensin-converting enzyme (ACE). Bradykinin is a potent vasodilator that elevates blood pressure and increases the risk of cardiovascular disease (CVD) and is inhibited by ACE. Preclinical and clinical research has demonstrated that a balanced diet has substantial benefits for human health, including cardioprotective properties. For example, a diet rich in fruits and vegetables can help prevent cardiovascular disease. Peptides, polyphenols, vitamins, oligosaccharides, and fatty acids are bioactive substances derived from plants with cardioprotective qualities and the capacity to promote cardiac health. In addition, numerous studies indicate that plant extracts may aid in blood pressure regulation. Studies have shown that grape polyphenols inhibit the oxidation of low-density lipoprotein (LDL), which is one of the processes associated with the prevention of atherosclerosis. Reduced oxidative stress influences the cellular redox state by reducing inflammation, altering endothelial function, delaying cell senescence by activating new proteins, preventing platelet aggregation, and reducing blood pressure. A preliminary study examined the bioactive properties of glucan. Adding 6% -glucan to pasta for 30 days reduced the prevalence of LDL cholesterol and cardiovascular disease (CVD), according to the results.

The capacity of bioactive peptides derived from fermented buckwheat sprouts to reduce blood pressure in hypertensive rodents was evaluated. According to the researchers, a single dose of six bioactive peptides (DVWY, FQ, FDART, VAE, WTFR, and VVG) reduced the blood pressure of hypertensive rodents. Separate studies examined the ACE-inhibiting potential of chia seed peptides. The results demonstrated that bioactive peptides extracted from chia via sequential Alcalase-Flavorzyme proteolysis exhibited significant ACE-inhibiting properties. Toufektsian et al. examined the cardioprotective effects of anthocyanin-rich maize (20% seed diet) after eight weeks of consumption in male Wistar rats. Anthocyanins may have cardioprotective properties, as rats fed a high anthocyanin diet had substantially ($p=0.01$) fewer infarcts, coronary occlusions, and instances of reperfusion than rats fed a low anthocyanin diet. One of the most significant polyphenols in olive oil, hydroxytyrosol, can reduce the risk of heart disease and function as a cardioprotective agent, according to a review study (Doğanlar, 2018).

2.3.2 Anti-cancer

Dietary and lifestyle changes may result in epigenetic changes. At this point, finding and developing medications that prevent cancer by changing epigenetic features may be a critical initial step. The Hedgehog, Notch, and Wnt/β-catenin pathways are cancer stem cell self-renewal pathways that can be changed by natural compounds in the diet. Chlorogenic acid (CGA), a dietary polyphenol, has a role in a variety of critical biological processes. It is one of the most unique chemicals present in coffee and is generated by a variety of plant varieties. It promotes apoptosis in a range of human cancer cells, including leukemia and lung cancer cells, and hence acts as an anti-cancer agent, according to a study. The anti-cancer potential of eggplant extract was examined using a human stomach cancer cell line. The eggplant extract was shown to be substantially more harmful to human gastric cancer cells than to normal cell lines. With the help of eggplant, which has substantial antioxidant qualities and a high concentration of phenolic components, free radicals can be removed. As a result, eggplants can be utilized to prevent cancer. Second *in vitro* research looked at the anticancer potential of solasodine, a naturally occurring aglycone of eggplant glycoalkaloids. Solasodine lowers matrix metalloproteinase (MMP) synthesis, decreasing the lifetime of the A549 cancer cell line and preventing cell invasion (metastasis), according to the findings. Solasodine also inhibited PI3K/Akt signaling and lowered miR-21 expression. Overall, the findings suggest that solasodine may be utilized to prevent lung cancer. Polyphenols, for example, are plant-derived chemicals that can make cancer cells more sensitive to chemotherapy and radiation, stop carcinogenesis, and rectify harmful or disastrous epigenetic modifications in cancer cells. Although diet-based treatments that target epigenetic pathways are fascinating, translating these scientific discoveries into clinical or public health practices is currently difficult. According to one research article, BB and GEN are both anticancer agents. The serum from BB-treated animals prevented the development of mammospheres in MDA-MB-231 and MCF-7 cell lines. In GEN-treated mammospheres, PTEN, a tumor suppressor that stimulates the PI3K/Akt signaling pathway, was also dramatically elevated. Molecular process studies have revealed that GEN can boost PTEN expression while lowering AKT expression (Holmsen, 2010).

The second research used stem-like glioma cells (CD133+) to investigate the anticancer potential of Pomiferin, a flavonoid derived from the fruit of the Maclura pomifera plant. Pomiferin was discovered to limit CD133+ cells' capacity to form spheres and enter glioma neurospheres, as well as to decrease their quantity and survival. Pomiferin also reduced the expression of genes associated with multiple stemness in glioma cells, such as Nanog, Nestin, and BIM1. An investigation into the anti-cancer potential of pterostilbene, a blueberry compound (natural stilbene), revealed that it effectively reduced the production of cancer stem cells and their ability to spread in response to M2 TAMs (tumor-associated macrophages) stimulation by altering the signaling pathways associated with the epithelial-to-mesenchymal transition (EMT), specifically the NF-B/miR488 pathway. Fruit and vegetable flavonoids are thought to suppress the NF-B pathway, which is known to regulate inflammation, angiogenesis, cell proliferation, cell transformation, cancer cell invasion/metastasis and survival, and cell transformation. To prevent cancer, anthocyanins, flavanones, chalcones, isoflavones, and flavonols, which are found in legumes, fruits, spices, nuts, and vegetables, act as negative regulators of cell signaling pathways that mediate pro-inflammatory responses. As a result, they may be used as part of a cancer therapy strategy. Furthermore, the anti-cancer and anti-inflammatory activities of plant-derived terpenoids and carotenoids have been proven. They block NF-B signaling pathways, which are important in the development of inflammatory and tumorigenic illnesses. A recent ex vivo research looked at the prevention of colon cancer by bioactive chemicals (polyphenolics) obtained from raspberries. The study looked at how fermentation altered polyphenols in an ileal fluid (Doğanlar, 2018).

Microbial metabolites rose greatly during fermentation, but polyphenols reduced significantly. The post-raspberry ileal fermentate decreased DNA damage in the CCD 841 CoN normal cell line by 30% in an oxidative challenge COMET test. Raspberry polyphenols absorbed into the colon and their microbially produced catabolites, according to the study, may help prevent colorectal cancer. A previous study suggests that the 4 kDa buckwheat peptide has anti-cancer effects. The peptide suppressed cell growth in HepG2 (hepatoma), liver embryonic WRL 68, breast cancer (MCF-7), and L1210 (leukemia) cell lines with IC50 values of 33, 37, 25, and 4 M, respectively. Recent research looked at the anti-carcinogenic potential of quinoa-derived peptides. The investigations suggest that LWREGM, DKDYPK, RELGEWGI, DVYSPEAG, and IFQEYI are potential anti-cancer medicines due to their chemoprotective and antioxidant capabilities. In another, *in vitro* study using breast cancer cell lines Anthocyanin and anthocyanin-pyruvic acid adduct extracts were tested for anticancer effects on MDA-MB-231 and MCF7 cells. In breast cancer cell lines, these extracts were found to exhibit anti-cancer action by reducing cancer cell growth and acting as anti-invasive agents. MCF7 in comparison to MDA-MB-231. A recent study found that sorghum bran extract reduces the development of human colon cancer cells. This therapy, according to the study, blocked cell invasion and migration while triggering S-phase growth arrest and cell death. These effects were followed by changed expression of genes that are adaptive to the cell cycle, apoptosis, and metastasis when cancer cells were treated with a sorghum bran extract containing 70% ethanol and 5% citric acid. Recent studies have confirmed quercetin's anti-doxorubicin vascular toxicity as well as its influence on the anticancer drug's therapeutic cytotoxic profile in breast cancer cell lines. The IC50 value for doxorubicin alone in MCF-7 cells was 0.4 0.03 M; when coupled with quercetin, this value rose to 0.8 0.06 M. Furthermore, annexin-V/FITC labeling revealed that doxorubicin/quercetin treatment resulted in considerably fewer apoptotic cells than doxorubicin therapy alone. According to the study's findings, quercetin modifies doxorubicin-induced anti-breast cancer potential via cellular pharmacokinetic and pharmacodynamic properties, as well as its greater vascular protective capacity compared to doxorubicin (Holmsen, 2010).

2.3.3 Anti-diabetes

Diabetes is a metabolic illness characterized by a variety of symptoms and problems such as glucose intolerance, prolonged hyperglycemia, and inadequate metabolic energy storage. Because of insufficient insulin production and insulin receptor activity, the condition affects the anabolic and catabolic digestion of lipids, carbs, and proteins. It is an illness that is wreaking havoc on people's health and spreading around the world. According to the WHO, diabetes mellitus (DM) affects 415 million people globally. An aqueous extract of Rutgers Scarlet Lettuce (RSL) strong in polyphenols, according to Cheng et al.'s research, may have anti-diabetic characteristics. RSL extract (RSLE) at dosages of 100 or 300 mg/kg enhanced glucose metabolism in obese mice produced by a high-fat diet after 28 days of oral treatment. Chlorogenic acid and RSLE inhibited glucose synthesis in H4IIE rat hepatoma cells in vitro. *In vivo* and *in vitro* investigations revealed that RSLE has anti-diabetic effects. Other studies have revealed that bioactive peptide fractions produced from quinoa have anti-diabetic effects. In vitro, investigations show that three bioactive peptide portions of 11S seed storage globulin B inhibit -glucosidase, -amylase, and dipeptidyl peptidase IV (DPP-IV) enzymes. Another research looked into the possible health benefits and capacity to prevent diabetes of Chungkookjang (short-term fermented soybeans with *Bacillus licheniformis*). Chungkookjang has been proven to increase insulinotropic activity in the islets of type 2 diabetic rats because of the presence of isoflavonoid aglycones and short peptides. Furthermore, 3 kDa peptides with PPAR antagonist action were thought to improve insulinotropic function.

Anthocyanin's anti-diabetic activities were evaluated in a 24-week clinical research with 58 diabetic individuals. Diabetics were given two anthocyanin capsules containing 160 mg of anthocyanin extracted from blackcurrant and bilberry. The anthocyanin group exhibited considerably lower fasting plasma glucose and insulin resistance scores, as well as significantly greater levels of adiponectin and -hydroxybutyrate in the blood, as compared to the placebo group. A study with streptozotocin-induced diabetic rats proved the anti-diabetic activity of peptides produced from oats. Alkalase-digested oat seed protein hydrolysate produced bioactive peptides that inhibited -glucosidase. Furthermore, in STZ-induced diabetic animals, oat peptides decreased food intake, enhanced glycogenesis, increased insulin secretion, and had a hypoglycemic impact.

Walnut hydrolyzed peptides (WHPs), which were produced from the proteins of *Juglans mandshurica* fruits, inhibited enzyme activity and hence demonstrated anti-diabetic characteristics. In an *in vitro* investigation, 3-10 kDa peptides increased extracellular glucose absorption while decreasing -glucosidase activity in insulin-resistant HepG2 cells. WHPs boosted liver glucokinase, glycogen, and insulin synthesis by 69.54 percent, 76.19 percent, and 23.71 percent, respectively, to reduce fasting blood glucose levels by 64.82 percent during *in vivo* testing. According to a recent study, black currants (BC), which are high in anthocyanins, can lower hyperglycemia in CaCo-2 cells.

Anthocyanins found in British Columbia control sugar transporters, glucose absorption, and salivary -amylase, potentially lowering the risk of developing type 2 diabetes. By reducing -glucosidase activity, BC anthocyanins significantly decrease postprandial hyperglycemia, whereas other phenolics affect sugar transporters, glucose absorption, and other factors. The glucose-lowering ability of puerarin (isoflavones) was investigated using STZ-induced diabetic animal models. The puerarin group had considerably reduced blood sugar (hypoglycemic effects) and greater blood insulin levels after four weeks of treatment. Furthermore, the puerarin group did better on OGTTs and had fewer injured islet cells. Furthermore, this isoflavone has been found to raise p-AKT and p-GSK-3 levels in liver tissue while lowering uncoupling protein (UCP2) mRNA levels. Puerarin was thought to reduce hyperglycemia in STZ mice by protecting pancreatic beta-cells and improving liver function via AKT signaling activation and UCP2 inhibition (Doğanlar, S. 2018).

2.3.4 Gut health

Experts regard "gut health" to be a unique medical research target based on epidemiological and animal studies. It has sparked attention not just as a therapy for severe gastrointestinal issues, but also as a possible strategy for infection prevention and resistance. The gastrointestinal tract's complex epithelium barrier isolates the host's gut lumen from the internal environment. The gastrointestinal barrier encompasses the immune system (cytokines, immunoglobulin A), the microbiota, and the intestinal vascular, neurenteric, and endocrine systems, in addition to mucous layers and epithelial cells. The gastrointestinal microbiota can influence how lipids and cholesterol are digested by controlling specific metabolic signaling pathways linked with the characteristics of bile acids, which operate as signaling molecules. Bile acids work by attaching to a variety of cellular receptors, including the farnesoid X receptor (FXR) and the G-protein-coupled receptor 5 (TGR5). Primary bile acids increased FXR signaling, which affects several target genes involved in lipid and glucose metabolism, as well as gut innate immunity (Holmsen, 2010).

Plant-based foods, such as fruits and vegetables, should make up the bulk of a person's diet since they are high in fiber and carbs. They also include a high concentration of other molecules (phytochemicals) that have a variety of impacts on the gut flora. Changes in gene expression mediated by protein hydrolysates and essential amino acid sequences (peptides) may have a hypocholesterolemic impact by binding to neutral sterols and bile acids in the intestines and causing fecal content to be removed, in addition to polyphenolic compounds. Polyphenols have the potential to improve gastrointestinal health and avoid pathological diseases linked with increased intestinal permeability (IP), according to a new clinical trial including 51 participants. Changes in phase II methylation mechanisms and gastrointestinal microbial metabolism may be to blame for IP-induced polyphenol bioavailability. Furthermore, it was established that the biological response generated by polyphenols in the age-related degeneration of IP was most likely mediated by microbial metabolites. An earlier study discovered that oolong tea epigallocatechin compounds promoted the growth of the beneficial *Bifidobacterium* spp. and Lactobacillus/Enterococcus species while inhibiting the growth of the harmful Eubacterium, Clostridium, *C. histolyticum*, and Bacteroides-Prevovobacterium groups. These findings show that oolong tea extracts may be able to modify the gut microbiota and improve the host's health. In a fourteen-month dietary interventional research, phenolic compound-rich grape pomace extract changed the intestinal microbiota of animals. In the control group, grape pomace extract consumption reduced the incidence of *Clostridium* (cluster I) while boosting the population of beneficial Bifidobacteria. Another study used a weaned piglet model to assess the efficacy of grape seed proanthocyanidins on gastrointestinal microbiota and intestinal barrier integrity. Proanthocyanidin treatment resulted in improved growth parameters, a much lower incidence of diarrhea, decreased intestinal permeability, and improved mucosal morphology in pigs. Proanthocyanidins also boosted the expression of the occludin gene in the intestinal barrier and improved blood and mucosal antioxidant indices. Furthermore, proanthocyanidins boosted gut bacteria diversity by increasing the number of Clostridiaceae and lowering the number of Lactobacillaceae (Doğanlar, 2018).

In a previous study, it was found that a polyphenol-rich extract of cranberry reduced intestinal inflammation in rodents. Eight weeks of supplementation with a polyphenol-rich cranberry extract significantly improved the Akkermansia spp. population, which was associated with reductions in weight gain, intestinal inflammation, and visceral obesity, which were all exacerbated by a high-fat/high-sugar diet. Recent research indicates that the high phenolic content and dietary fiber in whole cereals can alter intestinal flora, which has positive health effects. It has been hypothesized that cereal-rich diets if processed appropriately, could be used to treat and possibly manage a variety of metabolic diseases. Foods contain polyphenolic compounds and dietary fiber, which microorganisms can use to produce beneficial metabolites such as short-chain fatty acids (SCFA) and phenolic acids. The dietary fiber and phenolic content of foods may therefore aid in the maintenance of intestinal microbiota-mediated regulation and the prevention of chronic diseases. Gut dysbiosis, or disruption of the gut microbiota, has been linked to a variety of host health problems. In addition, SCFAs transported in the blood serve as key regulatory molecules in the mitogen-activated protein kinases (MAPK), AMP-activated protein kinases, and peroxisome proliferator-activated receptor gamma (PPAR) signaling pathways, thereby promoting lipid biosynthesis and gluconeogenesis. A recent study discovered that combining inulin with adzuki bean extract, which is rich in polyphenols, increased the concentration of propionic acid in rodents' caecum while diminishing the amount of ammonia nitrogen. The findings supported the use of dietary fiber and phenolic compounds/extracts as food components in gut-health-influencing functional foods.

Researchers examined the effects of *Porphyra tenera* (PT) seaweed meal on gastrointestinal health and microbiota modulation in mice fed dextran sodium sulfate (DSS) with PT extracts, DSS-fed mice, and control mice. The PT extract significantly altered the intestinal microbiota of mice and increased the number of Clostridium_XIVb species and other expected metabolic activities. They concluded that by altering the composition and activity of the gastrointestinal microbiota of mice, PT extract may reduce the inflammation of colitis caused by DSS. In turn, metabolites such as secondary bile acid, phenolic metabolites, and SCFAs generated by the intestinal microbiota in response to cereals and other plant-based diets are advantageous to host physiology. It is believed that microbial metabolites are the most important regulators of human metabolism. Through some signaling pathways, they may also influence oxidative stress and inflammation, as well as glucose and lipid metabolism, appetite, energy, and oxidative stress (Holmsen, 2010).

2.3.5 Lipids profile regulation

According to published scientific studies, bioactive compounds can maintain lipid homeostasis. Important phenolic and flavonoid components can influence lipid equilibria. The antihyperlipidemic effects of eggplant peel-derived anthocyanins on rodents were investigated. According to the study, anthocyanins from eggplant peel significantly increased high-density lipoprotein cholesterol (HDL-C) levels in rats while lowering levels of very low-density lipoprotein (VLDL), low-density lipoprotein cholesterol (LDL-C), serum total cholesterol, and triglycerides. It has been suggested that ingesting more anthocyanins from eggplant may aid in the prevention of obesity by lowering blood cholesterol and triglyceride levels while increasing high-density lipoprotein cholesterol levels. Natural plant sterols, particularly phytosterols, are used as functional food additives due to a vast array of potential health benefits, including the ability to lower blood cholesterol and reduce intestinal cholesterol absorption. Due to their structural similarity, these compounds inhibit the solubilization of cholesterol in the intestines, thereby reducing cholesterol absorption. By competing with LDL cholesterol, these compounds can also reduce the amount of low-density lipoprotein cholesterol (LDL) in blood vessels. According to research published in peer-reviewed journals, individuals who require a substantial reduction in plasma levels of low-density lipoprotein (LDL) cholesterol can safely consume up to 3 g of phytosterols per day (Doğanlar, 2018).

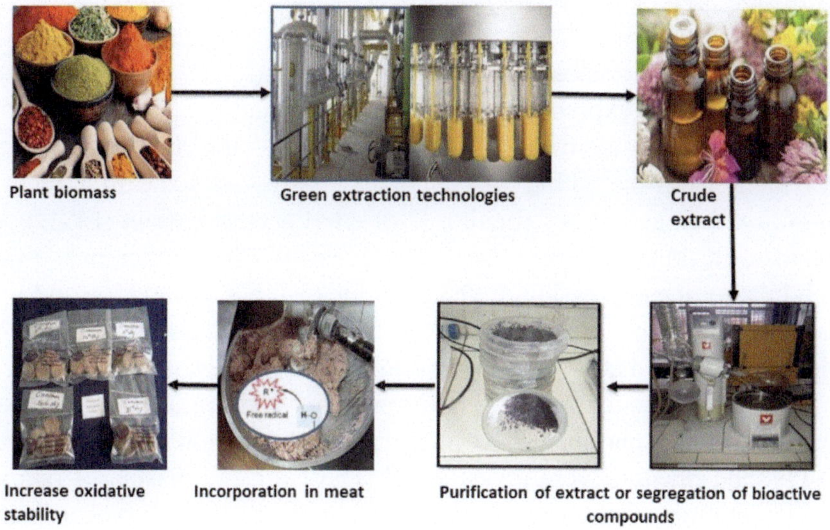

Figure 2.5. Recent developments in bioactive compounds.

Source: Awad AM, Kumar P, Ismail-Fitry MR, Jusoh S, Ab Aziz MF, Sazili AQ. Green Extraction of Bioactive Compounds from Plant Biomass and Their Application in Meat as Natural Antioxidant. *Antioxidants*. 2021; 10(9):1465. https://doi.org/10.3390/antiox10091465

Flavonoids can accelerate the decomposition and elimination of cholesterol via neutral sterols and bile acids, according to some studies. In male Sprague Dawley rats, oral administration of flavonoids extracted from unripe Musa paradisiaca fruits at a dose of 1 mg/100 g body weight (BW) produced a significant hypolipidemic effect. In addition, hepatic and fecal neutral sterols and bile acids increased significantly, indicating a faster rate of cholesterol oxidation. Patent rights were granted to Galvez for the biological properties of lunasin for cholesterol reduction. By ingesting sufficient quantities of lunasin, the method described in the patent reduces total cholesterol and LDL cholesterol levels while maintaining an individual's overall lipid profile. In a rodent model, bread containing *Moringa oleifera* (MO) leaf powder showed promise as an anti-hyperlipidemic agent. The most beneficial MO-fortified bread for lowering total cholesterol, triglycerides, low-density lipoprotein, and very low-density lipoprotein were those containing 10% and 15% MO. MO leaves also contain β-sitosterol, a bioactive molecule with well-established cholesterol-lowering properties, which may have contributed to the decrease in plasma cholesterol levels in rodents administered a high-fat diet.

Consuming a pomegranate peel extract strong in polyphenols can help rodents with elevated cholesterol levels caused by high-fat diets. In rodent colons and visceral adipose tissue, they discovered that pomegranate peel extract reduced the expression of inflammatory markers associated with a high-fat diet. A study confirmed the cholesterol-lowering efficacy of defatted fruit residue from the edible oil plant Idesia polycarpa. The significant *in vitro* antioxidant activity of fraction D of ethyl acetate extract (EF-D) was found to be attributable to the polyphenols present in this extract. In addition, by modulating the expression of genes implicated in inflammation and lipid metabolism, EF-D had a significant lipid-lowering effect on hepatic steatosis in oleic acid-derived HepG2 cells, resulting in increased antioxidant activity and decreased liver damage. On plasma lipids and antioxidant defenses, the effects of phospholipids, tocopherols, sterols, coenzymes CoQ10 and (Co) Q9, and phenols were investigated. The liver and plasma of rats fed a diet containing 20% fat and rapeseed oil were examined. According to the researchers, lower levels of triglycerides and cholesterol were observed depending on how the micronutrients were provided (Holmsen, 2010).

2.3.6 Oxidative stress and brain health

For health, metabolic systems must work properly. These pathways generate reactive oxygen species (ROS) can be dangerous. When electrons are and released from the mitochondrial electron transport chain by incomplete oxygen reactions, ROS are mostly formed. Carotenoids are advantageous to health due to their increased physiological effectiveness as pro-vitamins and antioxidative capabilities, particularly in the elimination of singlet oxygen. Furthermore, they can lower the risk of getting illnesses such as some cancers, which are a big worldwide health problem. According to a new study, terpenes and other bioactive substances, such as phytosterols, may benefit in the treatment of neuroinflammation and other neurodegenerative illnesses. Furthermore, there is evidence that certain fruit and vegetable extracts, particularly grape juice, contain significant quantities of antioxidants that boost brain function by mitigating the effects of oxidative stress and aging. The beneficial benefits of the phytoestrogen SDG (secoisolariciresinol diglucoside) on *Caenorhabditis elegans* health. SDG can improve the longevity of C. elegans by up to 22.0%, reduce the lethality of heat and oxidative stress, postpone the degeneration of the dopaminergic brain system caused by 6-OHDA, and diminish the toxicity of A protein, according to researchers. Resveratrol, a polyphenol found in red wine, is a well-studied biomolecule that may improve brain function by preventing beta-amyloid protein aggregation. Recent research looked at the effect of MO isothiocyanate on LPS-stimulated RAW 264.7 macrophages in a mouse model of Parkinson's disease (PD).

The MO bioactive molecule, which is produced when myrosinase is hydrolyzed, has a considerable influence on the pathways that govern oxidative stress, inflammation, and apoptosis, according to the study. According to one study, the *Ganoderma lucidum* fungus may be beneficial to rats who eat 3 g/kg of ethanol every day. The mice's sad and anxious profiles, as well as their potential for spontaneous horizontal exploration and memory impairment induced by excessive ethanol intake, were all recovered by an aqueous extract of *G. lucidum* (AEGl) (Doğanlar, 2018). As a result, AEGl can induce positive behavioral modifications in a model of mental, cognitive, and motor disorders. Additionally, it can be used as a dietary supplement to help alleviate problems caused by excessive alcohol consumption. Memory loss and dementia can be induced by an ischemic stroke in Wistar rats; however, a phytosome composed of mixed mulberry and ginger extract (PMG) may be able to mitigate these effects.

PMG elevated glutathione peroxidase (GSH-Px), superoxide dismutase (SOD), catalase (CAT), neuron density, and signal transduction via the extracellular signal-regulated kinase (ERK) pathway, whereas it decreased acetylcholinesterase, interleukin-6 (IL-6), and malondialdehyde (MDA). Overall, PMG was found to substantially enhance memory. As a consequence of oxidative stress, lipids, proteins, and nucleic acids can be degraded by reactive oxygen species (ROS) produced within cells. Eating bananas, which are rich in antioxidative bioactive components, can prevent the human body from a variety of conditions resulting from oxidative stress. Human embryonic kidney 293T cells (HEK 293T cells) were subjected to an *in vitro* experiment to determine whether the flavonoids present in lemon seeds (FLS) could mitigate the oxidative damage caused by peroxide.

The researchers discovered that treating damaged cells with 150 g/mL flavonoids increased their survival rate by up to 76.2% and prevented peroxide-induced apoptosis. According to Western blot and quantitative polymerase chain reaction, FLS increased the protein and mRNA expressions of GSH1, SOD1, SOD2, CAT, and GSH-Px in HEK 293T cells where H2O2 caused oxidative damage. FLS contained hesperidin, epicatechin, caffeic acid, vitexin, quercetin, and gallocatechin, according to the high-performance liquid chromatography study. Citrus fruits and beverages are the primary sources of flavanones in the diet. Isoprostane, a biomarker of oxidative stress, decreased in triathletes who consumed an aronia-citrus beverage high in flavanones for 21 weeks. A recent study described how beta-cryptoxanthin binds to retinoic acid receptors to inhibit angiogenesis. Carotenoid-rich diets may therefore be effective in the treatment of angiogenic pathological disorders such as moist macular degeneration and age-related cancer formation (Holmsen, 2010).

2.3.7 Influenza vaccine

Vaccination is critical for preventing seasonal influenza infections in high-risk individuals such as babies, the elderly, and people with impaired immune systems. There are now three types of influenza vaccinations available: recombinant hemagglutinin virus vaccine, inactivated virus vaccine, and live attenuated virus vaccine. Antigens from both circulating IBV lineages (quadrivalent vaccinations) or IAV H1N1 and H2N3 in addition to the circulating IBV lineage (trivalent vaccines) are included in these vaccines.

The inactivated influenza vaccine (IIV), which includes pure HA protein and is approved for use in people over the age of six months, is a subunit vaccine. It is injected either intramuscularly or intradermally. IIV are further categorized as trivalent (TIV) and quadrivalent (QIV) based on their antigenic content. TIV includes the IAV subtypes H1N1 and H3N1, as well as the predicted IBV. QIV, on the other hand, contains two IBV lineages that provide greater resistance to IBV in addition to the H1N1 and H3N2 IAV strains. IIV will result in the formation of strain-specific serum IgG. The temperature-sensitive attenuated live viruses in the live attenuated influenza vaccine (LAIV), also known as QIV, are delivered intranasally.

Only the nasal and lower respiratory tracts are affected by the attenuated LAIV virus. LAIV injection produces strain-specific IgG in the blood, IgA in the mucosa, and a T lymphocyte-mediated immune response. Furthermore, one study found that the efficiency of LAIV did not align well with the antigens of the recommended vaccinations. According to a study, LAIV helps avoid influenza infection in babies aged 6 to 59 months. It is, however, only available to anyone aged 2 to 49 (Doğanlar, S. 2018).

FluBlok, a recombinant HA virus immunization, is only available to those between the ages of 18 and 49. The pure HA protein in this vaccine is generated in insect cells using baculovirus vectors. Despite its tight age limits, the recombinant HA vaccine, which has a shorter production time than IIV and LAIV, would be effective during a pandemic emergency. A vaccination's efficacy and effectiveness influence its ability to defend itself. Seasonal vaccination protection for children, the elderly, pregnant women, and persons with impaired immune systems is more effective when the vaccine strain matches the ambient pressure, according to studies. This protection is also strengthened when the circulatory pressure falls within a predefined range. However, these vaccinations have several drawbacks, including the fact that they do not provide the best protection against pandemic influenza viruses, the lengthy vaccine development process, insufficient immunocompromised patient prevention, and the yearly immunization requirement.

To address the problem, many techniques have been developed. To lessen IIV and LAIV's reliance on embryonic eggs, some approved influenza vaccines, notably Flublock (HA recombinant vaccine), have been made outside of eggs. Flucelvax has received rights for IIV, a next-generation vaccine that does not require embryonated eggs (which is still in phase II clinical trials), viral vectors, DNA vaccines, and virus-like particles (VLPs). IIV uses cell culture techniques to develop the viral strain in mammalian cells (Holmsen, 2010).

Vaccine development is the most time-consuming due to the antigenic variation of the influenza virus, and yearly immunization is necessary. The virus, which is expected to circulate one month before the vaccine is created, is then synthesized, inactivated, purified, and disseminated in vials. The propagated viral projections are the most important phase. To choose the HA, antigenic mapping, a computerized study of HAI test data that offers a statistical framework for qualitative antigenic data interpretation is employed. However, vaccine manufacture and distribution were not finished in time for the 2009 pandemic. As a result, vaccine seed viruses (chosen based on epidemiological data) against numerous subtypes were stored to avoid a similar predicament during the next pandemic.

Researchers are striving to produce a universal influenza vaccine in light of the requirement for yearly vaccines and the insufficiency of current protection. We anticipate that the development of a widely cross-protective influenza vaccine will eventually replace yearly seasonal flu vaccinations and give protection against future outbreaks of a new influenza virus. The universal vaccines rely on the M2 proteins' substantially conserved outer domain and the hemagglutinin stem conserved epitopes (Doğanlar, 2018).

Despite the low immunogenicity of M2 generated by natural infection, M2 has been shown to defend against a range of influenza viruses. In a

VLP or recombinant vaccination, the M2 protein is normally generated by fusing the M2 gene or tandem repetitions of the M2 ectodomain (M2e) to a carrier protein or polymer. Furthermore, two methods are being investigated to address the insufficient protection of immunocompromised patients: increasing the antigen dosage to increase the immune response or adding an adjuvant. The second technique involves the use of oil-in-water adjuvants such as MF59 and AS03, which are generally well-tolerated and have been demonstrated to improve immunological responses to IIV, notably in the elderly and newborns (aged 6 months) (Holmsen, 2010).

Chapter

3 SEED STORAGE PROTEINS AS SOURCES OF BIOACTIVE PEPTIDES

CONTENTS

The plant seeds are high in nutrients. They contain less lignin and cellulose and more proteins, carbohydrates, and lipids than other plant parts. The majority of the proteins in the seeds are in storage form, are biochemically well characterized, and show no enzymatic activity. Because of their prevalence and adaptability, scientists have attempted to capitalize on the various characteristics of seed storage proteins (SSP). In addition to meeting the essential amino acid requirements for human and animal nutrition, SSP is a source of a variety of bioactive peptides with antihypertensive, antioxidative, immunomodulating, opioid, and other activities. These peptides, which are kept secret within the latent SSP primary structure, may be released during *in vivo* digestion or food processing. Numerous bioactive peptides share structural characteristics such as short residue length, hydrophobic amino acid residues, and the presence of Arg, Lys, and Pro. Specific bioactive peptides are expected to have a wide range of applications in the future as functional ingredients in foods that promote good health. Fewer bioactive peptides from plant proteins have been studied than from animal protein sources, including milk (Utsumi, 2011).

3.1 GENERAL DESCRIPTION OF SEED STORAGE PROTEINS

Seeds are the primary plant tissues harvested by humans. They are a significant source of dietary proteins, with protein contents ranging from 10% (in cereals) to 40% (in some legumes and oilseeds). Seeds contain two types of proteins: storage proteins and housekeeping proteins. Storage proteins are plentiful. Housekeeping proteins are in charge of maintaining normal cell metabolism. Seed proteins are classified into three types: storage proteins, structural proteins, and biologically active proteins. The biologically active proteins, such as enzymes, lectins, and enzyme inhibitors, are part of the cellular defense mechanism, whereas structural proteins, such as cell walls, help to maintain the structure. Proteins with biological and structural activity are scarce. Non-enzyme proteins in high concentrations serve as a reservoir for the amino acids required for seed germination and seedling growth. These proteins are known as SSP. SSP accounts for the majority of human protein consumption and also accounts for the majority of proteins in grains. SSP is the main protein body in seed protein storage vacuoles (PSV). Mature seeds have densely packed SSP deposits that fill the PSV (Wang, 2021).

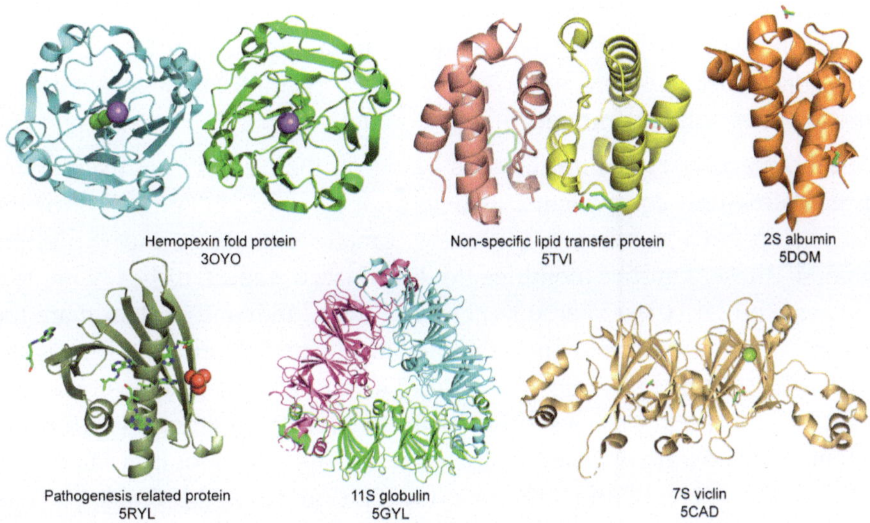

Figure 3.1. Structural features of seed storage proteins.

Source: Jain A. Seed Storage Protein, Functional Diversity and Association with Allergy. *Allergies*. 2023; 3(1):25-38. https://doi.org/10.3390/allergies3010003SSPs account for a significant portion of human protein consumption, either directly or indirectly. The proteins found in edible seeds are vital to the world's population. Cereals, legumes, and nuts can be used as sources of seed storage proteins, which are bioactive peptides, to meet their dietary protein needs and requirements. No single type of seed contains the entire range of essential amino acids required by humans and domestic animals. For example, legume proteins lack the S-containing amino acids Met and Cys, whereas cereal proteins lack Lys and Trp.

Because of its abundance and economic importance, SSP was one of the first plant proteins to be identified. In-depth studies on SSP began after Osborne (1924) classified it into the following groups based on extraction and solubility. Glutamins (the most difficult to solubilize), albumins (water extractable), globulins (extractable in diluted salt solutions), prolamins (extractable in aqueous alcohol), and globulins (extractable in diluted salt solutions) are among these. Albumins can also be extracted with water. Monocot storage proteins include prolamins and glutelins, whereas dicot storage proteins include albumins and globulins. When used as SSP, they typically contain high levels of Asn, Gln, and Arg or Pro, which act as a nitrogen reserve for the seedling. The majority of SSPs have high molecular weights and are poorly soluble in water. This may be important in the deposition and higher-order aggregation of storage proteins due to the osmotic conditions present in the drying and ingesting seeds. Alfalfa, Amaranthus, castor beans, Chenopodium, mung beans, mustard, pea, and many other oil-containing seeds are just a few examples of dicot seeds that contain water-soluble amounts of SSP (Utsumi, 2011).

SSPs are classified into families based on their amino acid sequences and 3D structural components, as determined by an understanding of the relationship between protein structure, genetic encoding, and evolution. Some of these families combine to form large superfamilies with low sequence similarity but shared structural traits. Cupin and prolamin are the two most common protein superfamilies discovered. Kreis (1985) was the first to identify the prolamin superfamily using an eight-residue conserved skeleton. The prolamin superfamily includes low molecular mass sulfur-rich proteins (such as dicot seed 2S albumins, trypsin and -amylase inhibitors, puroindolines, cereal grain softness proteins, soybean hydrophobic protein, and non-specific lipid transfer proteins) and SSP prolamines. Cupin proteins are distinguished from other proteins by their "barrel" or "jelly-roll" like structure. SSP globulins (7S vicillin-like globulins and 11S legumin-like globulins) are members of the cupin superfamily.

3.1.1 Cereals

In comparison to legume seeds, cereal grains contain relatively little protein (10-12% by dry weight). SSPs account for roughly half of the total proteins in mature cereal grains. Lys, Thr, and Trp are typically absent. Prolamins and glutelins account for approximately 80-90% of cereal SSP, with albumins and globulins accounting for the remaining 10-20%. Prolamins and glutelins are found in roughly equal amounts in wheat, barley, and rye. In comparison to glutelins, prolamins are abundant in sorghum and maize. Glutamins are overproduced in oats in comparison to prolamins. It is unique because glutelin accounts for a large portion of the protein in rice. Unlike glutelins, which form large disulfide-bonded aggregates, cereal prolamins only exist as monomers or small aggregates (Wang, 2021).

3.1.2 Legumes

Legume seeds have a high concentration of salt-soluble globulins (7S and 11S of the cupin superfamily). Sulfur-containing amino acids such as Cys, Met, and Trp are frequently deficient in legume proteins. They contain a lot of Arg, glutamic, and aspartic acidamides. The legume globulins are named legumin (11S) and vicillin (7S). Vicillin contains dicarboxylic acids, their amides, and trace amounts of Met, Cys, and Trp but no disulfide bonds (Utsumi, S. 2011).

3.1.3. Oilseeds like Brassica

Cottonseed (*Gossypium hirstum L.*), , rapeseed/canola (*Brassica napus L., B. rapa L. and B. juncea*), linseed (*Linum usitatissimum L.)* sunflower seed (*Helianthus annus L. var marcocarpus DC*), soybean *(Glycine max (L.) Merr.*), and peanut (*Arachis hypogae L.*). In contrast to legumes, the majority of oilseed proteins are used as animal feed after oil extraction. Soybean protein, which is commercially available in a variety of forms, is by far the most commonly used seed protein in foods. The health benefits of consuming soy protein have also prompted more scientific research into the protein. Except for soybean (11S glycinin and 7S conglycinin) and flaxseed (12S linin and 2S conlinin), the major SSPs in oilseeds are 11S globulin and 2S albumin.

3.2 BIOACTIVE PEPTIDES

In addition to meeting amino acid requirements, food proteins, including seed proteins, can produce bioactive peptides with a variety of physiological and health benefits. These "bioactive peptides" are defined as specific protein fragments that have a positive effect on bodily processes or conditions and may affect health. They can be released during GI digestion, hydrolysis by proteolytic microorganisms, or by proteolytic enzymes derived from microorganisms or plants [26]. They are inactive within the parent protein's sequence. Once released, bioactive peptides perform a variety of functions *in vitro* and *in vivo*, including antihypertensive, antimicrobial, antioxidative, antithrombotic, and immunomodulatory effects. As a result, depending on their amino acid sequence, bioactive peptides may have an effect on the major bodily systems after oral administration, including the nervous, immune, cardiovascular, and digestive systems (Wang, 2021).

Food-derived bioactive peptides typically contain 2 to 9 amino acids, but this range can be extended to 3 to 20 amino acid residues or more. For example, lunasin, a food-derived peptide with anticancer activity, has 43 amino acids and a molecular weight of 5400 Da. To carry out biological functions, these peptides must either pass through the intestinal epithelium and enter the bloodstream, or they must bind directly to specific epithelial cell surface receptor sites. Small peptides account for a sizable portion of dietary nitrogen absorbed by the body. The small intestine is the primary site of absorption for these protein digestion by-products, which are cleaved into amino acids and small peptides by a variety of *peptidases* secreted by the human GI tract. Small (2 to 3 amino acids) and large (10 to 51 amino acids) peptides produced by the body's diet can be effectively absorbed by the intestine and have biological effects at the tissue level. The intestinal epithelium may have a peptide transport system that actively transports peptides through the intestinal membrane while subjected to an H+ gradient, according to the findings of electrophysiological studies conducted in the 1970s and 1980s. This mechanism, however, only transports di- and tri-peptides. Oligopeptides with more than four residues can also pass through the intestinal epithelium via pinocytosis or paracellular channels, depending on their molecular size and hydrophobicity. After absorption, serum peptidase in the intestinal tract can further hydrolyze the peptide bonds. As a result, maintaining the physiological effects of bioactive peptides after oral or intravenous administration necessitates resistance to peptidase degradation.

Animal protein sources, such as milk, have been the primary source of bioactive peptides isolated and studied. The overwhelming evidence, on the other hand, points to SSP as physiologically significant bioactive peptide precursor. Numerous *in vitro* and *in vivo* studies, which are discussed in greater detail below, have confirmed this (Utsumi, 2011).

3.3 SSP-DERIVED BIOACTIVE PEPTIDES AND PHYSIOLOGICAL FUNCTIONS

3.3.1 Angiotensin I-Converting Enzyme (ACE) Inhibiting Peptides and Antihypertensive Effect

High blood pressure, also known as hypertension, is a serious health problem that affects 50% of people over the age of 55 in many developed countries. It increases the risk of kidney disease, congestive heart failure, stroke, and coronary heart disease. In humans, the systolic blood pressure (SBP) is measured during the active contraction phase of the cardiac cycle, whereas the diastolic blood pressure (DBP) is measured during the relaxation phase of the cardiac cycle.

Blood pressure control is heavily reliant on ACE. It is a zinc metalloprotease, a dipeptidyl carboxypeptidase, and a component of the rennin-angiotensin system (RAS). It catalyzes carboxy-terminal dipeptide hydrolysis from oligopeptide substrates. Other RAS constituents, such as

ACE, coexist locally in these tissues with blood vessels, kidneys, adrenal glands, the heart, and the brain. ACE concentrations are high on epithelial surfaces such as the intestinal, choroids plexus, and placental brush borders, affecting electrolyte and fluid balance at these fluid membrane interfaces. By catalyzing the degradation of bradykinin, a nonapeptide that lowers blood pressure, as well as the conversion of the decapeptide angiotensin I into the potent vasoconstrictor angiotensin II, ACE raises blood pressure. Angiotensin II's central effects include many outcomes that contribute to an increase in blood pressure. As a result, inhibiting ACE has a broad antihypertensive effect (Wang, L. 2021).

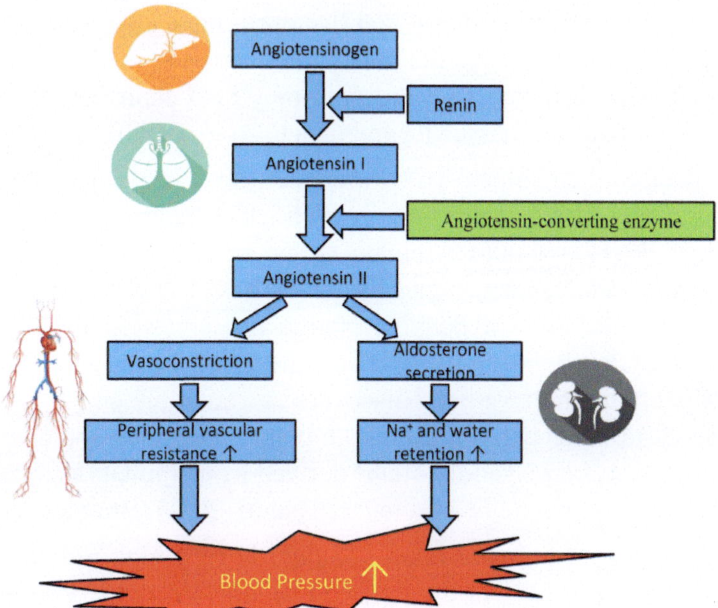

Figure 3.2. Effects of angiotensin-converting enzyme (ACE) inhibition.

Source: Jiang Q, Chen Q, Zhang T, Liu M, Duan S, Sun X. The Antihypertensive Effects and Potential Molecular Mechanism of Microalgal Angiotensin I-Converting Enzyme Inhibitor-Like Peptides: A Mini Review. *International Journal of Molecular Sciences*. 2021; 22(8):4068. https://doi.org/10.3390/ijms22084068

The N and C domains of ACE each have an active site containing the amino acids His-Glu-XX-His. It is thought that the His residues aid in Zn binding. The glutamate residue binds the activated water molecule and catalyzes the reaction by initiating a nucleophilic attack on the substrate's vulnerable peptide bond. The C domain is the most important angiotensin 1-converting site among these. The naturally occurring pentapeptide Glu-LysTrp-Ala-Pro and the nonapeptide Glu-Trp-Pro-Arg-Pro-Gln-Ile-Pro-Pro were the first competitive inhibitors of ACE isolated from the venom of the *Bothrops jararaca* snake. Structure-activity correlation studies on these peptide analogs revealed that their carboxy-terminal tripeptide residues (Trp-Ala-Pro) dominate competitive binding to the active site of ACE. Captopril, a highly effective ACE inhibitory (ACEI), was developed based on the antihypertensive mechanism of these peptides. The Zn atom in the active site of ACE forms a strong bond with the captopril sulphydryl group, which is coupled to the dipeptide Ala-Pro and inhibits enzyme activity. Many other ACE inhibitors, such as enalapril and lisinopril, have also been developed. According to Cheung (1980), the most effective C-terminal amino acids in peptides are Trp, Tyr, Phe, and Pro, and they also contribute the most to substrate binding at the active site and thus to ACE activity inhibition.

The database contained 168 dipeptides and 140 tripeptides. ACEI dipeptides prefer amino acid residues with thick or hydrophobic side chains, according to Wu et al. (2006). Aromatic amino acids have been discovered to be the most advantageous residues for the carboxyl terminus in tripeptides, while amino acids with a positive charge are preferred for the middle position and hydrophobic amino acids are preferred for the amino terminus. Many ACE inhibitors have been discovered through the enzyme-catalyzed hydrolysis of proteins derived from both plant and animal sources, such as enzymatic digests of SSP and proteins derived from animals such as bovine casein, human casein, and fish proteins (Utsumi, 2011). Brassica sp.

Canola/rapeseed seeds (*Brassica napus*) have a high protein content, accounting for 20 to 25% of the dry seed weight. In comparison, the meal left over after oil extraction contains approximately 40% protein. Four potent ACEI peptides isolated from the Subtilisin® digest of rapeseed protein are Ile-Tyr, ArgIle-Tyr, Val-Trp, and Val-Trp-Ile-Ser. Arg-Ile-Tyr and Val-Trp-Ile-Ser had IC50 (50% inhibitory concentration) values of 28 and 30 M (0.013 and 0.016 mg/mL, respectively). Each of these peptides reduced the blood pressure of spontaneously hypertensive rats (SHR) after oral administration. The Arg-Ile-Tyr peptide has also been shown to reduce blood pressure in older rats, where ACE is less effective, implying that ACE inhibition may not be the mechanism by which Arg-Ile-Tyr reduces blood pressure (Wang, 2021).

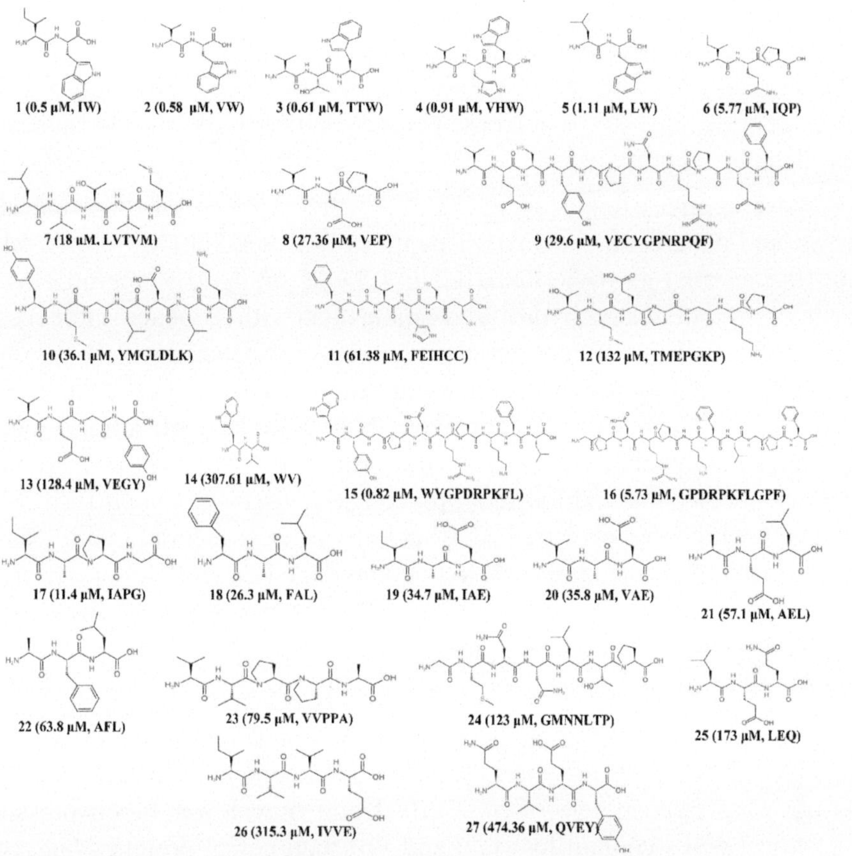

Figure 3.3. The structural formulas of microalgal ACE-inhibitory peptides.

Source: Jiang Q, Chen Q, Zhang T, Liu M, Duan S, Sun X. The Antihypertensive Effects and Potential Molecular Mechanism of Microalgal Angiotensin I-Converting Enzyme Inhibitor-Like Peptides: A Mini Review. *International Journal of Molecular Sciences*. 2021; 22(8):4068. https://doi.org/10.3390/ijms22084068

Wu et al. (2006) discovered ACEI activity in canola protein hydrolyzed with *Alcalase®* with an IC50 of 0.19 mg protein/mL. Furthermore, these peptides were resistant to GI protease incubation. The majority (87%) of the molecules in the fraction with the strongest ACE inhibition (IC50 of 0.014 mg protein/mL) were 870 Da peptides. Aromatic amino acids contributed up to 37% of the amino acids in these peptides. SSP in canola protein is a good source of ACEI peptides after being hydrolyzed by food-grade microbial enzymes, as demonstrated by the ACEI activity of *Alcalase®* hydrolyzed rapeseed protein and defatted canola meal. Wu and colleagues discovered that the IC50 values for ACEI activity in hydrolyzed heat-treated and untreated canola meals were 0.027 to 0.028 mg/mL and 0.035 to 0.044 mg/mL, respectively. This process may contribute to the production of more ACEI peptides due to the increased susceptibility to hydrolysis caused by heat-denaturing proteins.

These peptides retained their activity after *in vitro* GI digestion with pepsin and pancreatin. Two ACEI peptides, Val-Ser-Val (IC50: 0.15 M or 0.0510-3mg/mL) and Phe-Leu (IC50: 1.33 M or 0.3910-3mg/mL), were purified from *Alcalase*® hydrolyzed canola meal with sequence references to napin and cruciferin. When canola protein was hydrolyzed with *Alcalase*®, the IC50 values for isolated napin and cruciferin were 0.015, 0.035, and 0.029 mg/mL, respectively. However, there is a lack of solid evidence regarding the specific canola SSP that produces ACEI peptides. Another Brassica species that exhibit this property is *Brassica carinata* (Ethiopian mustard), which demonstrated ACEI activity after sequential hydrolysis by trypsin, chymotrypsin, and carboxypeptidase A (IC50 of 0.338 mg protein/mL) (Utsumi, 2011).

Buckwheat

The protein content of pseudocereal buckwheat (*Fagopyrum esculentum Moench*) is 12%. Buckwheat, among other potential health benefits, has a strong hypocholesterolemic effect and may lower blood pressure. Buckwheat SSPs contain less Glu and Pro than cereal proteins, but more Asp, Arg, and Lys. Buckwheat seeds' main storage protein is a 13S globulin with a structure similar to legumin.

ACE inhibitors with an IC50 value of 0.043 mg protein/mL were produced by thermolysin®-assisted enzymatic hydrolysis of buckwheat globulin. According to Li and colleagues' findings, *in vitro* digestion of buckwheat flour with GI proteases can increase the digest's ACEI activity (IC50 of 0.14 mg protein/mL compared to 3 mg/mL for intact buckwheat). When fed to SHR, buckwheat digest reduced SBP. The tripeptides TyrGln-Tyr and Pro-Ser-Tyr have been identified as the primary ACE inhibitors in buckwheat digest. Scientists isolated and purified an ACEI peptide from buckwheat protein extract without the use of enzymes. The tripeptide Gly-Pro-Pro was discovered to be an ACE inhibitor, with an IC50 value of 6.25×10-3mg protein/mL. Proline was the C-terminal amino acid of the ACEI peptides found in buckwheat (Wang, 2021).

Chickpea

The crude protein content of chickpeas (*Cicer arietinum L.*) ranges from 12.6% to 30.5% [60]. Chickpea is regarded as a good source of vegetable proteins due to its well-balanced amino acid composition, high protein bioavailability, and generally low level of associated antinutritional factors. The primary storage protein is legumin, which accounts for 97% of all globulins and 64% of the total protein in chickpea seeds. According to Pedroche and colleagues, the hydrolysate produced by breaking down chickpea protein isolate with Alcalase has ACEI activity. Four ACEI peptides were isolated from the hydrolysate. Two of them competitively inhibited ACE, while the other two inhibited it in an uncompetitive way. Yust et al. (2003) isolated six ACEI peptides with IC50 values ranging from 0.011 to 0.021 mg/mL using Alcalase® to treat isolated legumin. Although hydrophobic amino acids were abundant, methionine was the most common amino acid in all of these peptides. Yust et al. discovered one ACEI peptide made entirely of Met among these. Using ACEI peptides Sardine muscle, human αs1 casein, and β-*lactoglobulin* were also used to extract met residues. There are no *in vivo* studies on the ACEI peptides produced by chickpea proteins.

The field pea

The protein content of field pea (*Pisum sativum L.*) varies from 15.6 to 32.5% depending on cultivar and growing season conditions. Pea protein is primarily composed of globulins (11S and 7S) and heterogeneous albumins (larger PA2 albumin and sulfur-rich PA1 albumin). The globulin fraction contains approximately 80% of all proteins found in mature pea seeds. According to several *in vitro* studies, pea proteins may act as an ACEI peptide precursor. Vermeirssen et al. (2003) investigated the ACEI activity of a hydrolysate of isolated pea protein, which had an IC50 value of 0.076 mg/mL after GI digestion simulation. Through in silico GI digestion of the vicilin and albumin PA2 proteins, the ability of pea SSP to release a variety of potent ACEI peptides was discovered. Vermeirssen et al. fermented pea proteins with *Lactobacillus helveticus* and *Saccharomyces cerevisiae* and then subjected them to simulated GI digestion in a different study. When the IC50 values of fermented and unfermented pea proteins were compared at the end of GI digestion, the former had the lower IC50 value, indicating that GI digestion is the primary factor controlling the release of ACEI peptides from pea proteins. Microbial proteases may prevent GI proteases from hydrolyzing pea proteins, allowing them to release more potent ACEI peptides. Vermeirssen et al. fractionated the GI digest of pea protein isolates using ultrafiltration and reverse-phase chromatography to increase ACEI activity. They discovered that the ACEI activity was multiplied by four and that all of the pea peptides had similar ACE inhibitory activity. Pea protein isolates digested with GI proteases significantly reduced mean arterial blood pressure in SHR by 44.4 mmHg after receiving 50 mg of protein/kg of body weight intravenously. Only a small percentage of the ACEI peptides in the pea protein digest, which are the source of small intestinal peptidases, were transported through Caco-2 cells. Because Caco-2 cells are tighter than mammalian intestinal cells, sufficient absorption of these peptides may occur under actual *in vivo* conditions (Utsumi, 2011).

Maize

The main proteins in maize seeds are zein and glutelin, which account for roughly 60% and 30% of the total protein in maize (*Zea mays*) seeds, respectively. While glutelin has a nutritionally balanced amino acid composition, zein is deficient in Lys and Trp, as well as to a lesser extent in Arg, His, and Met. As a result, it is claimed that maize protein has a low nutritional value. Based on their solubility, zeins are classified into four major groups ((α, β, γ and δ). Among these proteins, α-zein has the highest level of hydrophobicity. Thermolysin® hydrolysis produced the ACEI tripeptides Leu-Arg-Pro, Leu-Ser-Pro, and Leu-Gln-Pro, with IC50 values of 0.27, 1.7, and 1.9 M (110, 590, and 745 mg/mL, respectively). The hypotensive activity of Leu-ArgPro on SHR was demonstrated by a significant drop in blood pressure (15 mmHg) after a 30 mg/kg intravenous injection. SHR's blood pressure was also reduced when they were given unpurified Thermolysin® hydrolysate of α-zein orally. In a study, scientists used Thermolysin® to denature urea-denatured α-zein and convert it into small peptides. The ACEI activity (IC50) of this Thermolysin® digest of α-zein was 24.5 μg/mL (0.024 mg/mL). 36 ACEI peptides were isolated from this hydrolysate, including 5 dipeptides, 14 tripeptides, 9 tetrapeptides, 5 pentapeptides, and 3 hexapeptides. Despite its low nutritional value, zein is a valuable source of bioactive peptides due to its ability to release ACEI peptides when hydrolyzed by enzymes (Wang, 2021).

Mung beans

Mung beans (*Phaseolus radiatus* L.) have well-known anti-inflammatory, anti-tumor, cholesterol-lowering, detoxifying, and diuretic properties due to their high protein content, which ranges from 20 to 33%. The seeds are ground and processed into flour, sprouts, mature seeds, and cooked whole or split beans. Mung bean protein has been found to be a good source of ACEI peptides when hydrolyzed *in vitro* with the protease Alcalase®. An increase in ACEI activity of the mung bean hydrolysates after treatment with GI proteases *in vitro* indicated the presence of pro-drug type peptides in the hydrolysate. SBP was significantly reduced when this hydrolysate was administered orally to SHR. The ability of mung bean protein to produce ACEI peptides is explained by the protein's high concentration of hydrophobic amino acids. ACEI activity has been demonstrated in tripeptides with hydrophobic amino acids at their C and N terminals. Lys-Asp-Tyr-Arg-Leu, Val-Thr-ProAla-Leu-Arg, and Lys-Leu-Pro-Ala-Gly-Thr-Leu-Phe were isolated from mung bean protein hydrolysate treated with *Alcalase*® and have IC50 values of 26.5 μM, 82.4 μM, and 13.4 μM (0.02, 0.06, and 0.01 mg/mL, respectively). The hydrophobic amino acids Leu and Phe, which are located at the C-terminal positions of Lys-Asp-Tyr-Arg-Leu and Lys-Leu-Pro-AlaGly-Thr-Leu-Phe, respectively, may influence the ACEI activity of these peptides. Furthermore, the peptide Lys-Leu-Pro-Ala-Gly-Thr-Leu-Phe shares the C-terminal dipeptide with the *in vitro* potent ACEIs Tyr-Gly-Leu-Phe and Tyr-Leu-Leu-Phe derived from μ-lactalbumin. According to research, the positive charge on the arg and lys -amino groups in the C-terminus of peptides contributes to their ACEI potency, explaining the ACEI of Val-Thr-Pro-Ala-Leu-Arg. The mung bean SSP that serves as an ACEI peptide precursor has yet to be identified.

Sesame

Sesame (Sesamum indicum L.) seeds contain 40-50% protein, are fat-free, and high in S-amino acids. Sesame is used in confections, personal care products, and juice and soft drink fortification, whereas protein hydrolysates are used in savory applications such as soups, sauces, gravies, snacks, and meats. In addition to oil, sesame contains protein and unique lignans. Ochi et al. (1995) isolated three ACEI-active peptides from a Thermolysin® hydrolysate of sesame seed proteins: Met-Leu-Pro-Ala-Tyr, Val-Leu-Tyr-Arg-Asp-Gly, and Ileu-Val-Tyr. When given orally at a dose of 100 mg/kg, each peptide had an antihypertensive effect in SHR. Some scientists isolated and identified six ACEI peptides from commercially available sesame peptide powder. Ileu-Val-Tyr, Val-Ileu-Tyr, Leu-Val-Tyr, Leu-Gln-Pro, Leu-Lys-Tyr, and Leu-Ser-Ala are the names of these compounds. The most pronounced ACEI activity is found in tripeptides with Tyr and Pro at the C-terminus and Leu at the N-terminus. Cheung et al. (1980) reported substrate specificity for ACE, which may lend support to these findings. It has also been shown that peptides made from sesame peptide powder reduce SBP in SHR (Utsumi, 2011).

Soybeans

The two main proteins found in soybeans (*Glycine max L. Merr.*) are glycinin (11S) and -conglycinin (7S), though other minor proteins are also present. ACEI activity has been demonstrated in mouse models and chemical assays for several soy protein-derived peptides. Gly, nonpolar (Ile or Val), aromatic (Tyr or Phe), polar (Gln), positively charged (His), or negatively charged (Asp) amino acid residues have been found at the carboxyl and amino terminals of ACEI-active soy peptides. Val was the most frequently found residue in ACEI soy peptides. The following amino acids were found in decreasing order: Leu, Phe, Asp, Pro, Gln, Gly, Ala, Ile, Asn, Glu, His, Thr, Arg, Met, and Lys. His-His-Leu (HHL), a soybean peptide fraction isolated from fermented soybean paste, was found to have *in vitro* ACEI activity with an IC50 value of 2.2 mg/mL in one study. Furthermore, the synthetic tripeptide HHL decreased SBP in the SHR and significantly reduced ACE activity in the aorta.

The significant hypotensive effect of ACEI peptides on SHR at a dose of 100 mg/kg body weight/day in soy protein treated with *Alcalase®* demonstrated their *in vivo* bioactivity. However, when compared to the clinical hypotensive drug captopril, the ACEI activity of soy peptide was found to be significantly lower. The ACEI peptides Asp-Leu-Pro and Asp-Gly had IC50 values of 4.8 μM (1.82×10^{-3} mg/mL) and 12.3 μM (2.5×10^{-3} mg/mL), respectively when produced *in vitro* by the hydrolysis of soy protein by Alcalase®. The four ACEI peptides produced by peptidic digestion of soy proteins are Phe-Phe-Leu (IC50: 37 μM or 0.01 mg/mL), Ile-Tyr-Leu-Leu (IC50: 42 μM or 0.02 mg/mL), and Ile-Ala (IC50: 153 μM or 0.03 mg/mL). These peptides were tested for their ability to lower SHR blood pressure when administered orally at a dose of 2g/kg body weight. SHR's mean SBP dropped to 17.6 mmHg in 2 hours after oral administration, and this antihypertensive effect lasted for 6 hours. According to Lo and Chan (2005), the sequential *in vitro* digestion of soy isolates with pepsin and pancreatin produced peptides with ACEI activity, indicating that soy proteins have the potential to release peptides with such activity under physiological conditions (Wang, 2021).

However, further pancreatin hydrolysis reduced the ACEI activity of the peptides in the peptic hydrolysate. The digestion of soy hydrolysate and fermentation products such as natto and tempeh with a variety of endoproteases revealed that ACEI peptides were primarily derived from glycinin, a commonly expressed soy protein, and were found primarily in the *pronase*, kidney membrane proteases, and plasma proteases breaks down of fermented soy products. -conglycinin, a soy protein, was discovered to be resistant to tryptic proteolysis, acid hydrolysis, plasma proteases, and kidney membrane proteases. It also resisted acid hydrolysis. According to Lo and Chan, isolated soy proteins may be able to release ACEI peptides when exposed to conditions that mimic the upper GI tract. Because pancreatin digestion may reduce ACEI activity, studies simulating only the upper GI tract are insufficient to predict the potential of soy proteins as precursors of ACEI peptides. Despite extensive research into soy protein's ability to produce ACEI peptides, more clinical research is needed to demonstrate that these peptides are released and functional in a physiological setting.

Sunflower

Defatted sunflower seed meal (*Helianthus ananus L.*) contains approximately 30% protein that can be used as a food ingredient. Megias et al. (2005) isolated an ACEI peptide using a sunflower seed protein isolate pepsin and pancreatin digest. The peptide had the amino acid sequence Phe-Val-Asn-Pro-Gln-Ala-Gly-Ser and was discovered to be a fragment of helianthinin (11S globulin) with an IC50 of 5.7 x 10-3 mg/mL. Hydrolyzing isolated sunflower protein with Alcalase® and Flavourzyme® yielded ACEI peptides. The ACEI fractions of the hydrolysates were rich in Asp, Ser, Gly, Thr, and Val residues, in contrast to the original protein hydrolysate; however, some fractions were also rich in Ala, Met, Ile, Leu, or Phe residues, supporting the idea that ACEI peptides are rich in hydrophobic amino acids. Flavourzyme®-derived fractions had the lowest IC50 value (0.0810-3 mg/mL), which was significantly lower than the ACEI peptides released by *in vitro* GI digestion of sunflower protein, as reported by Megias. There is no evidence that sunflower peptides have ACEI activity (Utsumi, 2011).

Wheat

Wheat (*Triticumaestivum L.*) grain has a low protein content, ranging from 9 to 16% by dry weight. Wheat grain proteins are essential in processed foods such as bread, biscuits, breakfast cereals, and pasta. Wheat storage proteins that are alcohol soluble are gliadins and glutelins. Wheat germ (WG), a byproduct of the flour milling process, is an excellent source of vitamins, minerals, dietary fiber, proteins, and calories. Defatted wheat germ contains more than 30% protein and is especially high in Lys, Met, and Thr, three essential amino acids that are often lacking in cereal grains. *In vitro* inhibition of ACE was demonstrated using WG hydrolysate produced by alkaline protease hydrolysis of *Bacillus licheniformis*. Sixteen peptides with IC50 values less than 20 μM and amino acid sequences ranging from 2 to 7 were isolated from the hydrolysate. The major peptide with the highest ACEI activity was ile-val-tyr (IC50 0.48 μM or 0.2110-3 mg/mL). The ability of ACEI was further enhanced by simulating GI digestion of WG hydrolysate, indicating that GI protease activity results in the production of new ACEI peptides. Motoi and Kodama (1982) isolated the ACEI peptide from wheat gliadin hydrolysate produced by acid proteases. The amino acid sequence of this peptide was determined to be Ile-Ala-Pro (IC50 = 2.7 μM or 0.91 10-3 mg/mL). This peptide inhibited the hypertensive activity of angiotensin II when administered intravenously and significantly lowered blood pressure when administered intraperitoneally when tested using SHR (Wang, L. 2021).

Additional SSP-Derived Peptides with ACEI Activity

ACEI activity has been discovered in other SSP-derived peptides. They are made up of Thermolysin® hydrolyzed flaxseed meal, peanut, kidney, and red bean protein hydrolysates, and other ingredients. The amino acid residues and their arrangement in sequence determine the bioactivities of bioactive peptides. As a result, substituting amino acid residues in peptides can improve their biological activity. Antihypertensive transgenic rice with antihypertensive activity was created by introducing an antihypertensive ovokinin-like peptide structure from egg albumen into the hypervariable regions of glutelin, the main storage protein in rice.

3.3.2 Opioid Activity

In humans, the endogenous opioidergic system consists of opioid receptors and ligands, which are endogenous opioids with peptide or alkaloid structures that act by interacting with the receptors. Opioid receptors are found in mammals' immune systems, endocrine systems, and central and peripheral nervous systems. Specific organ tissues that resemble opioid activity include the spinal cord, adrenal gland, digestive tract, pituitary gland, and hypothalamus. The μ-, κ- and δ -type opioid receptors regulate emotional behavior as well as intestinal motility inhibition. While κ-type receptors are important for sedation and food intake, δ-type receptors are also associated with emotions and rewarding behavior. Endogenous opioid peptides include enkephalins, endorphins, and dynorphins. They have agonistic and antagonistic effects on opioid receptors in the human body and cause effects in all cells or tissues where opioids are known to be active. Opioid peptides affect both the GI system and the nervous system. Another mechanism for blood pressure reduction by bioactive peptides is opioid peptide fixation to blood vessel opioid receptors. According to a large body of research, opioid peptides have been implicated in regulating food intake and obesity in both humans and lab animals. Opioid receptor agonists promote a positive energy balance and obesity in rodent models of obesity, whereas antagonists at these receptors reduce food intake and body weight. It's worth noting that the strongest effects of opioid receptor antagonists on food intake and body weight are seen in obese animals or those fed a highly appealing diet. Opiates are well known for their interactions with monoamines in the central nervous system, which affects eating behavior. Endogenous opioid peptide release in the brain is also important in the sensory pleasure response to food. As a result, opioid antagonists that block opioid receptors may result in decreased food preferences and consumption. Endogenous opioids, in particular, regulate dietary fat preferences and sweet taste preferences. Endogenous opioid system abnormalities may be responsible for increasing the consumption of sweet and fatty foods, which contributes to overweight and obesity (Utsumi, 2011).

Exogenous opioid peptides are opioid peptides derived from food proteins that serve as exogenous supplements to the endogenous opiodergic system. It has been discovered that opioid peptides derived from food are essential for the human body. One explanation is that because food-derived opioids are released after intestinal enzymatic action, they are less susceptible to enzymatic degradation than endogenous opioid peptides. The other is that opioid peptides derived from food typically have lower activity than endogenous opioid peptides, making them less likely to result in opioid-related side effects such as dependence, tolerance, and addiction. The N-terminal sequence of exogenous and endogenous opioid peptides differs structurally. While several food-derived opioid peptides (e.g., Tyr-X-Phe, Tyr-X1-X2-Phe) have a Tyr residue at the amino-terminal, endogenous peptides share the same N-terminal sequence of Gly-Gly-Phe. Currently, several food-derived opioid peptides of animal and plant origin have been identified. It's worth noting that while opioid peptides derived from plant proteins are predominantly δ-receptor-selective, those derived from animal proteins are predominantly β-receptor selective. Following the discovery of the first food protein to bind an opioid receptor, several other opioid agonists and antagonists were discovered from milk proteins. This was a piece of bovine β casein. The opioid agonist and antagonist activities of alfalfa white protein concentrate, RuBisCo hydrolysates in spinach leaves, and several SSP have been studied among the opioids derived from plant proteins (Wang, 2021).

Gluten exorphins (GEs) are among the most extensively researched opioid peptides derived from SSP. These are opioid peptides extracted using enzymes from wheat gluten digestions. During the digestion of wheat proteins, peptides that interact with the opioid receptors in the gut are released. Zioudrou et al. (979) discovered that peptides in wheat gluten peptic hydrolysate have *in vitro* opioid activity similar to morphine. GEs are classified into three types based on their structural characteristics: GEA, GEB, and GEC. Both GEA5 (Gly-Tyr-Tyr-ProThr) and GEA4 (Gly-Tyr-Pro) have two members, as do GEB5 (Tyr-Gly-Gly-TrpLeu) and GEB4 (Tyr-Gly-Gly-Trp).

GEB5 is the most potent food-derived -opioid peptide found to date, with the highest *in vitro* activity as an agonist on both μ- and δ- receptors. GEB5 is released *in vivo* after gluten consumption and to successfully cross the intestinal barrier. It stimulates prolactin release in rats by acting on opioid receptors in nerves outside the blood-brain barrier. When Schusdziarra (1981) gave dogs digested gluten, he noticed that peripheral vein insulin and glucagon levels rose faster than when the same amount of undigested protein was given. According to additional research, the effect may be related to exorphins' activation of opiate receptors. Both oral and intravenous administration of GEA5 can stimulate insulin release in rats, according to Fukudome et al. (1995). Insulin has an appetite suppressant effect in humans. As a result, by stimulating insulin release, GEA5 may prevent excessive energy intake and the onset of obesity. The appetite-suppressing effects of insulin, however, are observed in lean subjects rather than obese subjects. There has been no research to show that GEs can exert comparable opioid activity in human subjects (Utsumi, 2011).

The same GEA5 peptide has been shown to suppress the body's natural pain-inhibitory system and to help mice learn and remember. GEs are also being studied for GI functions. Hydrolyzed gluten caused a longer intestinal transit time in healthy volunteers, which could be reversed by giving them naloxone, an opiate blocker. The longer transit time could be attributed to the reversible increase in plasma somatostatin-like activity caused by naloxone. However, the authors discovered no effect of GEs on volunteers' ability to control their appetite.

CCK, a critical physiological endocrine factor in appetite control, is produced in the brain and enteroendocrine cells. suppression of food intake is well documented. Soy hydrolysate consumption caused satiety in rats by sending satiety signals to their opioid and CCK receptors, according to Pupovac and Anderson (2002). Peptic hydrolysate of soyβ-conglycinin, a strong inducer of CCK release from intestinal mucosal cells, inhibits gastric emptying and food intake. The amino acid sequence Gly-Arg-Gly-Arg-Gly-Arg-Gly-ArgGly was discovered in the fragment of β-conglycinin that binds to the intestinal membrane of rats and stimulates CCK release. The appetite suppressant in β-conglycinin was found in the f51-63 region, and the numerous Arg residues in this region may have contributed to the manifestation of this effect.

Oryzatensin (Gly-Tyr-Pro-Met-Tyr-Pro-Leu-Pro-Arg), a bioactive peptide isolated from rice albumin tryptic digest, has been shown to contract longitudinal muscle strips in the guinea pig ileum. Oryzatensin demonstrated biphasic ileum contraction, which is characterized by a rapid contraction followed by a slower contraction. The latter was carried by the cholinergic nervous system. Even though oryzatensin had a low affinity for μ-opioid receptors, the slower contraction appeared to be associated with the apparent anti-opioid activity. The rapeseed-derived ACEI peptide Arg-Ile-Tyr reduced food intake and gastric emptying in mice after oral administration by promoting CCK release. This demonstrates that a peptide with a specific amino acid sequence can perform a wide range of bioactivities. Although the findings of these studies on the impact of SSP-derived peptides on oploid activity are promising, longer-term clinical trials are needed to determine whether these peptides can have similar effects in humans (Utsumi, 2011).

3.3.3. Antioxidant activity

In the human body, the metabolism of oxygen results in the constant formation of free radicals such as superoxide, nitric oxide, and hydroxyl, as well as other oxygen-derived species (hydrogen peroxide and hypochlorous acid). This is in addition to the oxidants already mentioned, which include air pollutants, ozone, nitrogen oxides, tobacco smoke, and vehicle exhaust.

Because they have an unpaired electron in their outer orbit, these reactive oxygen species (ROS) are extremely reactive. Superoxide and other molecules with unpaired electrons can generate free radicals, which can initiate chemical chain reactions. For example, in the presence of metal ions such as Fe2+ or Cu+ (the Fenton reaction when iron serves as the catalyst), the Haber-Weiss reaction can combine superoxide (O2) and H_2O_2 to produce the reactive hydroxy radical (OH). Such chain reactions contribute to lipid peroxidation, DNA damage, and protein degradation during oxidatively stressful events. The human body has several antioxidant defense systems, including enzymes like catalase and superoxide dismutase, to combat these ROS. These defenses, however, are not entirely effective. When ROS are produced in excess or antioxidative defenses are suppressed, numerous pathological events such as aging, cellular damage, and DNA degradation occur. Excessive free radical production has been linked to the onset of many diseases, including atherosclerosis, rheumatoid arthritis, and cancer. Antioxidants in the human diet are therefore of great interest as potential protective agents to assist the human body in reducing oxidative damage. Antioxidants can stop or inhibit oxidation by preventing the production of ROS or inactivating ROS.

Glutathione (GSH: Glu-CysGly), carnosine (β-Ala-L-His), anserine (β-Ala-3 methyl-L-His), and homocarnosine γ aminobutyryl-L-His are all naturally occurring antioxidant peptides with free radical scavenging activity in the human body. GSH, for example, scavenges free radicals by releasing its H atom and acting as a cosubstrate in the GSH peroxidase reduction of H_2O_2 and other hydroperoxides. Carnosine was one of the first His-containing peptides discovered in beef extract in 1900. It is one of the most prevalent (1-20 mM) nitrogenous compounds in the nonprotein portion of vertebrate skeletal muscle and some other tissues, such as the olfactory epithelium. Carnosine is also an O2 scavenger and a Cu chelator, both of which prevent the Haber-Weiss reaction in addition to scavenging OH (Wang, L. 2021).

Several seed protein hydrolysates have also produced antioxidant peptides. Six antioxidant peptide fragments are effective against lipid peroxidation were isolated from the soy-conglycininenzymatic digest. These peptides have 5 to 16 amino acid residues, with hydrophobic amino acids like Leu or Val at the N-terminus and His, Pro, and Tyr in the sequences. Chen et al. (1998) discovered that His-containing peptides can function as metal-ion chelators, active oxygen quenchers, and OH scavengers by starting with the smallest peptide in soy digest, LeuLeu-Pro-His-His. Pro-HisHis had the highest level of antioxidant activity of any peptide tested, with His and Pro both playing important roles in the peptide sequence's antioxidant activity. Oligopeptides isolated from soy hydrolysates, fermented soy, and papain-hydrolyzed soy protein have been found to have antioxidant activity. However, these findings contradict those of Pena-Ramos and Xiong (2002), who discovered that papain hydrolysis of soy protein does not result in antioxidant activity, implying that genotype influences antioxidant peptide release (Utsumi, 2011).

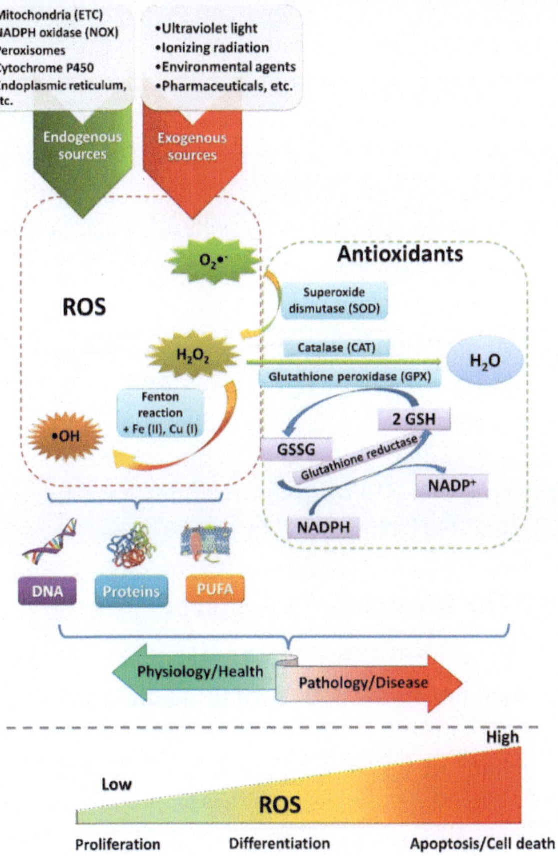

Figure 3.4. Reactive oxygen species damage biological molecules, increasing the risk of disease.

Source: Milkovic L, Cipak Gasparovic A, Cindric M, Mouthuy P-A, Zarkovic N. Short Overview of ROS as Cell Function Regulators and Their Implications in Therapy Concepts. *Cells*. 2019; 8(8):793. https://doi.org/10.3390/cells8080793.

Saito et al. (2003) created peptide libraries from antioxidative tripeptides isolated from soy proteins and assessed the antioxidative activity of the peptides using various *in vitro* techniques. Despite having a low reducing activity and a moderate peroxynitrite scavenging activity, the antioxidative tripeptide Tyr-His-Tyr demonstrated a strong synergistic effect with phenolic antioxidants. Tripeptides with Trp or Tyr at the C-terminus had high radical scavenging activity but low peroxynitrite scavenging activity, whereas tripeptides with Cys at the C-terminus had the highest peroxynitrite scavenging activity.

Wheat germ protein hydrolysates (WGPH) produced by *Alcalase*® catalyzed hydrolysis had antioxidant activity comparable to α-tocopherol in a linoleic acid emulsion system. DPPH, O2, and OH were just a few of the free radicals that WGPH could scavenge. *Alcalase*®-hydrolyzed zein has demonstrated antioxidant activity similar to or comparable to that of butylated hydroxyanisole (BHA), α-tocopherol, and ascorbate by acting as a metal ion chelator or a hydrogen donor, as well as a radical stabilizer to prevent lipid oxidation.

These results suggest that SSP contains antioxidant peptide sequences that can be released during enzymatic hydrolysis. An enzyme is used to catalyze the hydrolysis of proteins, which determines the hydrolysis conditions as well as the structure and amino acid sequence of the released peptides. All of these studies show that SSP-derived peptides have the potential to act as exogenous antioxidants (Utsumi, 2011).

3.3.4 Potential for lowering cholesterol

Dietary proteins have been shown to lower serum cholesterol levels. Soy protein, in particular, has been shown to have cholesterol-lowering properties in a variety of populations of children and renal patients. However, until recently, the mechanism underlying soy protein's ability to lower plasma cholesterol was unknown. There is evidence to suggest that the ability of these peptides to bind bile acids aids in the reduction of blood cholesterol levels. Bile acids are acidic steroids produced in the liver from cholesterol. They are actively reabsorbed from the terminal ileum and circulated through the enterohepatic circulation before being secreted into the duodenum to aid in digestion.

According to research on the hypocholesterolemic effect of soy proteins, a peptide with a high affinity for bile acids may be able to inhibit bile acid absorption in the ileum and stimulate the conversion of cholesterol to bile acids in the liver and plasma, lowering plasma cholesterol levels. These peptides may also lower blood cholesterol levels by decreasing cholesterol micellar solubility in small intestinal epithelial cells. Soy isoflavones may play a role in how effectively soy protein products lower cholesterol. Sirtori et al. (1997), on the other hand, demonstrated that soy protein products lacking in isoflavones can still significantly lower plasma cholesterol. Lovati et al. (2000) exposed HepG2 cells to synthetic peptides derived from specific -conglycinin sequences or peptides derived from *in vitro* digestion of a commercially available, isoflavone-deficient soy protein concentrate. The authors demonstrated that incubating Hep-G2 cells with a synthetic peptide increased LDL uptake and degradation, whereas incubating the cells with a digest of soy protein concentrate resulted in a significant upregulation of LDL-receptors.

According to the findings of this study, soy proteins produce low molecular weight peptides that can lower cholesterol. When given orally for two days at a dose of 50 mg/kg, leu-Pro-Tyr-Pro-Arg, a soy glycinin fragment, was shown to lower serum cholesterol in mice (reduced total and LDL cholesterol by 25.4% and 30.6%, respectively). This study's lack of an increase in fecal cholesterol and bile acid excretion suggests that bile acid binding may not be the mechanism causing hypocholesterolemic activity. However, studies show that peptides derived from soy glycinin can bind bile acids. When soy 11S globulin was compared to casein and 7S globulin, higher hypocholesterolemic activity was observed. The peptide Ile-Ala-Val-ProGly-Glu-Val-Ala binds to cholic and deoxycholic acids, causing a hypocholesterolemic effect. It was isolated from 11S globulin pepsin hydrolysate. Choi (2002) identified a possible bile acid linking peptide sequence (Val-Ala-Trp-Trp-Met-Tyr) in the acidic polypeptide A1a of the A1aB1b subunit, one of the five glycinin subunits (group I: A1aBIb, A1bB2, and A2B1a; group II: A3B4 and A5A4B3). By including the nucleotide sequence encoding this peptide in the DNA coding of A1a polypeptide, the ability of glycinin to bind bile acids was improved (Wang, 2021).

Soy protein peptic hydrolysate with phospholipids (SPHP) had significantly higher *in vitro* bile acid binding capacity than soy protein peptic hydrolysate without phospholipids (SPH). When compared to cholesterol micelles containing casein tryptic hydrolysate, cholesterol micelles containing SPHP and SPH reduced cholesterol uptake by Caco-2 cells significantly. *In vivo* rat feeding studies revealed that rats fed SPHP excreted significantly more total steroids in their feces than those fed SPH (Utsumi, 2011).

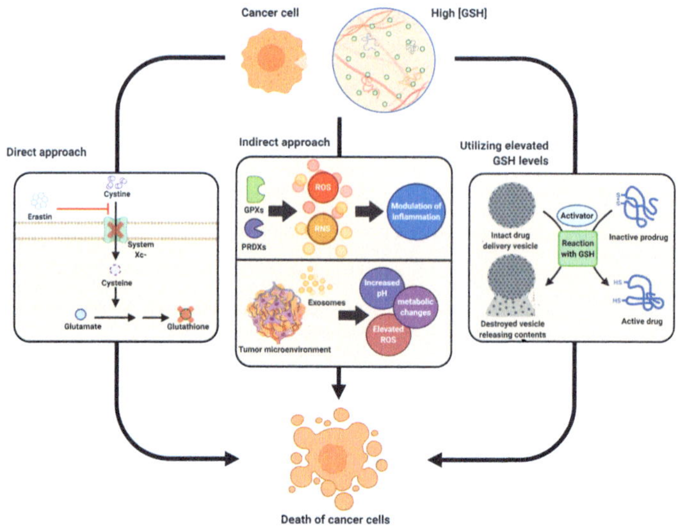

Figure 3.5. Reactive oxygen species (ROS) generation pathways and clearance by the tripeptide glutathione.

Source: Kennedy L, Sandhu JK, Harper M-E, Cuperlovic-Culf M. Role of Glutathione in Cancer: From Mechanisms to Therapies. *Biomolecules*. 2020; 10(10):1429. https://doi.org/10.3390/biom10101429

Cholesterol is converted into bile acids, which are then conjugated and expelled through the bile ducts. Bile acids are reabsorbed by the intestinal epithelium, which aids in the absorption of dietary cholesterol. Bile acid binding peptides derived from dietary proteins reduce bile acid re-absorption while increasing bile acid excretion with feces.

3.3.5. Modulation of the immune functions

The immune system protects the body from foreign microbes such as bacteria, viruses, and parasites through the actions of different cell types such as monocytes/macrophages, platelets, erythrocytes, polymorphonuclear leukocytes, and T and B lymphocytes. A plethora of bioactive peptides derived from plant proteins have been discovered to modulate immune function. Soy is a major source of plant protein in this group. Yoshikawa and colleagues isolated the peptide soymetide-13 (Met-Ile-Thr-Leu-Ala-Ile-Pro-Val-Asn-Lys-Pro-Gly-Arg) from the alpha subunit of β-conglycininβ, which stimulates phagocytosis in human polymorphonuclear leukocytes. It was discovered that the methionine residue at the amino terminus was required for this activity. The peptide His-Cys-Gln-Arg-Pro-Arg isolated from a tryptic digest of soy glycinin was also found to be immunostimulating, activating neutrophil phagocytosis and stimulating tumor necrosis factor (TNF) when given orally to mice (Wang, 2021).

The soybeantide agonist N-formyl-methionyl-leucyl-phenylalanine (fMLP) is the first food-derived peptide. fMLP, a synthetic version of the naturally occurring N-formylmethionyl peptide that is highly chemotactic for human neutrophils and macrophages, stimulates the immune system. Neutrophils and macrophages have specific fMLP receptors on their surface that control both the production of ROS by neutrophils and macrophages as well as neutrophil activity that stimulates phagocytosis. These actions result in a rapid response to bacterial infection because bacterial proteins contain an N-formylmethionine residue at their N-termini. Bacterial death is caused by phagocytosis and ROS-induced bactericidal effects. Soybeans have a low affinity for fMLP receptors. As a result, the immune system reacts to soy protein consumption in the same way it would to a bacterial infection. Soymetide-4, a tetrapeptide, is the shortest peptide that promotes phagocytosis. Soymetide-4 may not cause neutrophils to create ROS in a vitro setting because it has a lesser affinity for fMLP receptors than fMLP. . As a result, soymetide-4 could be used as an immunostimulant without causing inflammation. Orally administered soymetide-4 has also been shown to inhibit the anti-alopecia mechanism induced by the anti-cancer drug etoposide. Because of its short length, which makes it easier for the digestive system to absorb, soymetide-4 is more desirable as an immunostimulating peptide among the isolated soymetides (Utsumi, 2011).

Oryzatensin (Gly-Tyr-Pro-Met-Tyr-Pro-Leu-Pro-Arg), which is produced by tryptic digestion of rice albumin, has also been identified as an immunomodulatory peptide. Oryzatensin C-terminal fragments with comparable activity have also been reported. Oryzatensin induced *in vitro* phagocytosis and the production of superoxide anions in human leukocytes. Wheat gluten hydrolysate (hydrolyzed with amylase and protease) has been studied to see how it affects the immune systems of healthy human volunteers. After consuming 3 g of wheat gluten hydrolysate for 6 days, the natural killer (NK) cell activity in the test group significantly increased in comparison to the control group. NK cells play an important role in immune surveillance against viral infections and tumor growth. Thus, NK cell activation is beneficial for patients with cancer and autoimmune diseases, as well as the elderly, who typically have low levels of NK cell activity. Its immunostimulatory activity could be attributed to immunostimulating peptides. According to Clare et al., ACEI peptides can stimulate the immune system by inactivating bradykinin. Bradykinin stimulates macrophages, which promotes lymphocyte migration and lymphokine secretion.

3.3.6 Calmodulin binding activity

Among the many cellular enzymes controlled by the soluble protein calmodulin (CaM) in humans are cyclic nucleotide phosphodiesterase, Adenylyl cyclase, $Ca2^{+}$-$Mg2^{+}$ ATPase, nitric oxide synthase, calcineurin, and several protein kinases. CaM regulates cyclic nucleotide metabolism, Ca2+ transport, protein phosphorylation-dephosphorylation cascades, ion transport, cytoskeletal function, and cell proliferation via these enzymes. Protein kinase II and other CaM-dependent enzymes with high activity phosphorylate additional cellular proteins linked to chronic pathogenic diseases. As a result, any substance that can reduce CaM activity can inhibit these processes and halt disease spread. Peptides containing basic amino acids that give a net positive charge or peptides with hydrophobic surfaces have the potential to inhibit CaM since CaM is negatively charged and has an exposed hydrophobic surface in its active site. Many peptides with these properties have been found to bind to CaM and prevent CaM from activating phosphodiesterase. The strongest of these peptides was discovered to have an alpha-helical structure. Many SSP protein hydrolysates have been found to contain peptides with CaM binding activity (Wang, L. 2021).

The *in vitro* activity of CaM-dependent protein kinase II was decreased by pea proteins hydrolysate of Alcalase®-catalysis that was rich in positively charged amino acids like Lys and Arg and contained short peptides that bind CaM. The same enzyme has produced peptides from the hydrolysis of flaxseed protein isolate that can bind to CaM and inhibit CaM-dependent endothelial and neuronal nitric oxide synthase. There is no evidence to suggest that other SSPs could produce CaM-binding peptides or that these peptides would function in *vivo* (Utsumi, 2011).

3.3.7 Cancer Prevention

Peptides isolated from hydrolyzed SSP have been shown to have anti-cancer properties in both *in vitro* and *in vivo* studies. Soy protein isolates in diets, according to Xiao and colleagues, prevent tumorigenesis in rats by increasing somatostatin, a hormone known to inhibit the growth of colon cancer cells. This anticancer activity could be attributed in part to bioactive peptides produced during soy protein digestion. *In vitro,* cytotoxicity of defatted soy protein peptides against mouse monocyte-macrophage cell line (IC50 value of 0.16 mg/mL) was observed. This hydrolysate was purified further, yielding the nonapeptide X-Met-Leu-Pro-Ser-Tyr-Ser-Pro-Tyr. Lunasin, a chemopreventive bioactive peptide derived from soybean 2S albumin, has been shown in mice to reduce the occurrence of skin tumors and carcinogen-induced cell transformation. When applied exogenously, the lunasin peptide prevented the carcinogen-induced conversion of murine fibroblast cells to cancerous foci *in vitro*. Lunasin is a distinct, naturally occurring peptide with 43 amino acids and nine Asp residues at its carboxyl end, preceded by the cell adhesion motif Arg-Gly-Asp. It acts as a chemopreventive agent, possibly by binding to chromatin. Lunasin can prevent core histone acetylation by attaching to nonacetylated H3 and H4 histones. This mechanism is thought to be in charge of the anticancer properties of this chromatin-binding peptide. Lunasin has also been isolated from barley (*Hordeum vulgare L*), wheat, and amaranth (*Amaranthus hypochondriacus*) seed proteins. Lunasin from barley, which had been partially and crudely purified, suppressed colony formation in stably transfected mouse fibroblast cells. These fractions also reduced histone acetylation in mouse fibroblasts and human breast cells in the presence of the histone deacetylase inhibitor sodium butyrate. According to Jeong and colleagues, the bioactive lunasin can be extracted from the livers of rats fed a diet high in lunasin-rich wheat. This suggests that lunasin is still intact and biologically active after GI digestion. The glutelin fraction of amaranth seeds contained the most lunasin (3.0 g/g), which was discovered. Lunasin was also found in prolamin, globulin, albumin, and even popped amaranth seeds (Wang, 2021).

3.4. PREDICTION OF SSP-DERIVED PEPTIDE BIOACTIVITIES USING PEPTIDE DATABASES

Polymorphic gene families encode SSP in a controlled manner. Zein, for example, is encoded by 30 to 100 genes in maize. The primary determinants of bioactive peptides' bioactivities are thought to be their amino acid sequence and composition. For example, while most ACEI peptides have Pro residues at the C-terminus, most opioid peptides have Tyr and Phe residues. As a result, the primary structure of a protein's precursor proteins has a significant impact on its ability to produce bioactive peptides.

For predicting the bioactivities of peptides found within parent proteins, both information on isolated bioactive peptides and computer-aided databases are available. These peptide libraries contain information on peptides synthesized via genetic engineering or chemical synthesis.

Users can search for information on a specific peptide by comparing the experimental peptide mass and amino acid sequence to the molecular mass and amino acid sequence of peptides in the database. The precursor protein of a bioactive peptide can currently be predicted using many databases based on its known amino acid sequence. These databases also provide information on the genes that code for a specific peptide. The BIOPEP database classifies food proteins as potential sources of bioactive peptides. According to the database, the majority of food proteins contain antihypertensive peptide fragments. According to this database, rice prolamin and bovine β-casein are the best sources of antihypertensive peptide precursors. Bioactive fragments and their surroundings have been discovered to be both hydrophilic and primarily localized at random coil structures of proteins. According to Wang and Mejia (2004), database search using BIOPEP, soy proteins contain sequences with anti-amnestic, antihypertensive, antioxidant, antithrombotic, *dipeptidyl peptidase* IV inhibitory, immune-stimulating, and opioid activities. Soy proteins 7S and 11S were discovered to contain bioactive peptides, and the 2S protein has also shown promise. This matching resulted in the discovery of previously unknown opioid peptide sequences in soy proteins (Wang, 2021).

Dziuba (2004) used bioinformatics to analyze biologically active fragments and bonds of food proteins that are susceptible to the action of endopeptidases with known specificity. Wheat gliadins were found to be the most susceptible to the release of bioactive peptides. These peptides had antihypertensive, dipeptidyl peptidase inhibitory, opioid, and antioxidative effects after being released by the activity of chymotrypsin, *elastase*, ficin, or pepsin (Utsumi, 2011).

Chapter

4 CHEMISTRY AND BIOACTIVITY OF FOOD PHYTOALEXIN

CONTENTS

4.1 INTRODUCTION

Food phytoalexins are a group of biologically active compounds found in plants. These compounds can be found in seeds, leaves, stems, and roots of plants. They are known for their ability to protect plants from pathogens by killing them or inhibiting their growth. Food phytoalexins are considered as potential therapeutic agents for human diseases. For example, *Fusarium oxysporum f.sp. Lycopersici* (FOCL) causes postharvest rots in vegetables and fruits. The symptoms of postharvest rots include discoloration, softening of the fruit or vegetable, and rotting of the skin. Therefore, it is important to control this pathogen using various methods such as using fungicides or maintaining proper storage conditions. The bioactive components of food phytoalexins have been used as natural products to treat infections caused by bacteria and fungi including tuberculosis, *staphylococcal enterotoxemia*, Lyme disease, etc. The most common form of treatment for these infections is penicillin but some people cannot tolerate penicillin because it can cause nausea or diarrhea (Aluko, 2012).

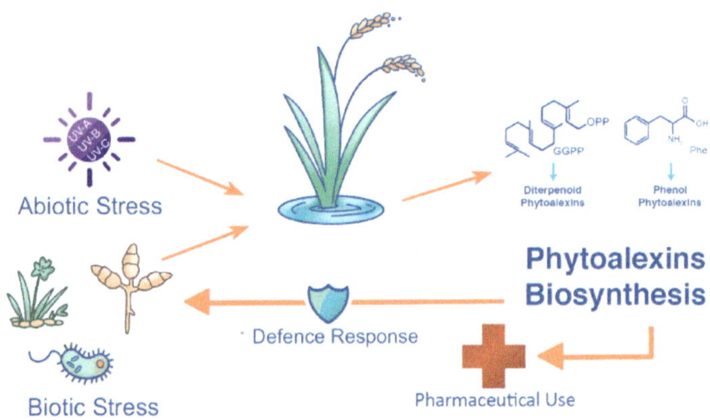

Figure 4.1. Representation of a sample rice phytoalexins.

Source: Valletta A, Iozia LM, Fattorini L, Leonelli F. Rice Phytoalexins: Half a Century of Amazing Discoveries; Part I: Distribution, Biosynthesis, Chemical Synthesis, and Biological Activities. *Plants*. 2023; 12(2):260. https://doi.org/10.3390/plants12020260

Phytoalexin can be found in many types of plants, including broccoli, red pepper, cabbage, celery, radish, potato, and tomato. These compounds are produced by plants when they are attacked by pathogens such as bacteria or fungi. This process is called an herbicidal mechanism. When phytoalexin is ingested by humans through foods or beverages that contain them, it has been shown to have a beneficial impact on human health by reducing the number of pathogenic microbes in the body. Some studies have shown that phytoalexin can help treat infections caused by viruses like influenza and colds. The ones found in food are generally considered to be safe, and some have even been proven to be beneficial to human health. The compound is produced by plants to defend themselves against pathogens and pests. It has also been shown to inhibit the growth of bacteria and fungi, as well as protect against infections caused by these organisms (Bizuneh, 2021).

Phytoalexins are more common in foods such as broccoli, cabbage, radishes, cauliflower, and Brussels sprouts because they contain higher concentrations of these compounds than non-food plants do. Cooking may increase their bioavailability (ability to be absorbed into the body), but they continue to exist even after cooking at temperatures below boiling or above freezing. Heat treatment does not affect phytoalexins' overall chemical structure or their ability to fight off infections. Because phytoalexin is found naturally in many different kinds of foods, there are many different types of phytoalexins depending on what plant was used for its production. For example, one type might come from cabbage while another type comes from broccoli! Different types may also have different effects on diverse animals and living organisms.

It's important to note that phytoalexin isn't just found in legumes—it's also found in fruit and vegetables like broccoli, cauliflower and spinach. This means you can add these foods to your diet if you want to get more phytoalexin into your system. Phytoalexin acts as an antimicrobial agent, meaning that it kills bacteria and viruses by disrupting their cell wall structure. It does this by interacting with enzymes known as topoisomerases. These enzymes are responsible for breaking down DNA molecules during replication, so when phytoalexin binds with them it prevents replication from occurring properly.

4.2 ROLE IN PLANTS

Phytoalexins act as antimicrobial agents, which means they kill microorganisms (such as bacteria and fungi) by interfering with their growth. Phytoalexins have been shown to have activity against a variety of Gram-positive and Gram-negative bacteria, including species that cause food spoilage and pathogenic infections. Phytoalexins can be divided into two main categories: plant lectins and glucosinolates. Plant lectins are glycoproteins that bind specifically to carbohydrate groups on the surface of certain bacteria, causing them to stick together and release toxins that kill them. Glucosinolates are sulfur-containing compounds, typically found in cruciferous vegetables such as broccoli and Brussels sprouts. These compounds have been shown to have antimicrobial activity against various strains of *S. typhimurium, S. mutans, S. sanguinis, S. gordonii, A. hydrophila, Pichia kodakaraensis, Candida albicans (including C. glabrata), Coccidioides immitis, Histoplasma capsulatum* and *Blastomyces dermatitidis* (Aluko, 2012).

Phytoaxelins help plants to protect themselves from pathogens and insects, as well as other hostile environmental conditions. They do this by defending against the attack of pathogens and insects, as well as other hostile environmental conditions. These compounds can be found in the roots and leaves of plants, and they have been shown to have some antimicrobial activity against bacteria, fungi, and viruses.

Phytoaxelins are most concentrated in seeds, fruits, leaves, flowers, and stems. Phytoalexins act as a type of natural antibiotic because they inhibit the growth of harmful organisms through a variety of mechanisms including interference with protein synthesis or inhibition of nucleic acid synthesis (such as DNA replication).

In addition to their antimicrobial properties, phytoalexins also have antioxidant properties that can reduce oxidative damage from free radicals caused by oxygen radicals or ultraviolet light exposure. They also may protect plants from disease-causing organisms such as bacteria or fungi by causing them to die out before they have time to reproduce themselves within the plant itself. They prevent pathogenic organisms from growing on the surface of plants by making the cell walls less permeable to water and nutrients. For phytoalexins to be effective against pathogens (like fungal or bacterial infections), they must be present at high concentrations within the plant tissues. This means that it's very important for farmers who grow these crops to harvest them quickly before any pathogens have time to develop resistance against them (Bizuneh,. 2021).

In general, they help protect plants from damage caused by UV light, oxygen, and other environmental stressors. The compound carvacrol is particularly effective at preventing damage caused by UVA rays. Some phytoalexins have also been found to have anti-inflammatory effects. This makes them useful in treating a variety of human conditions including arthritis, asthma, and allergies.

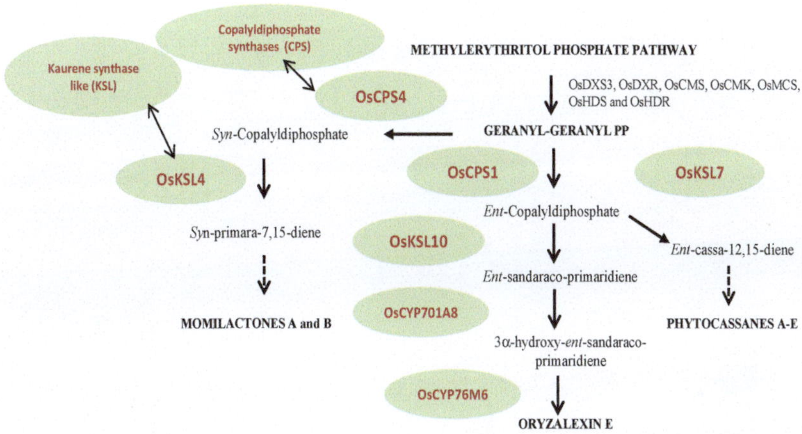

Figure 4.2. Deciphering the Role of Phytoalexins in Plant-Microorganism Interactions and Human Health.

Source: Jeandet P, Hébrard C, Deville MA, et al. Deciphering the role of phytoalexins in plant-microorganism interactions and human health. *Molecules*. 2014;19(11):18033-18056. Published 2014 Nov 5. doi:10.3390/molecules191118033

4.3 TYPES OF PHYTOALEXINS IN FOOD: CHEMICAL STRUCTURES AND SOURCES

The chemical structure of these compounds varies depending on their origin, but most phytoalexins have a wide range of biological activities. The term phytoalexin refers to any compound found in plants with antibiotic or antifungal properties. These compounds can be derived from the entire plant or just specific parts of it such as leaves, stems, roots, flowers, or seeds. Phytoalexins can also be produced by bacteria during fermentation processes. Phytoalexin content in food depends on many factors including climate conditions such as temperature and humidity levels; soil type; crop species; cultivar type; stage in the growth cycle and storage conditions. In addition, it also depends on how much time has passed since harvest or when harvested. There are two types of phytoalexins: indolamines and flavonoids. Indolamines are found in a variety of plant foods, including broccoli, Brussels sprouts, cabbage, cauliflower, and radishes. Flavonoids are found in soybeans and spinach (Aluko, 2012).

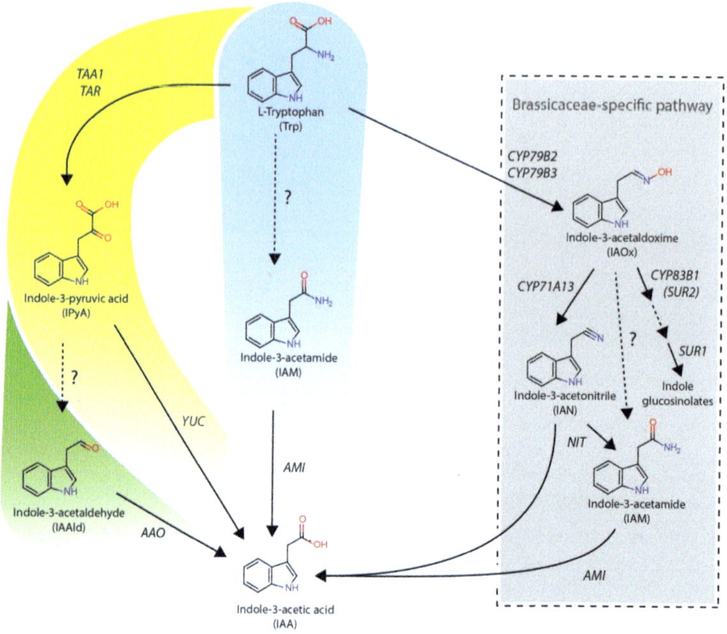

Figure 4.3. Tryptophan metabolism to indoleamines and auxin.

Source: Sánchez-Parra B, Frerigmann H, Alonso M-MP, Loba VC, Jost R, Hentrich M, Pollmann S. Characterization of Four Bifunctional Plant IAM/PAM-Amidohydrolases Capable of Contributing to Auxin Biosynthesis. *Plants*. 2014; 3(3):324-347. https://doi.org/10.3390/plants3030324

Indolamine phytoalexins bind to the gut wall and inhibit the growth of bacteria that cause diarrhea. They also reduce inflammation and increase insulin sensitivity. Indolamine phytoalexins can be used to treat inflammatory bowel diseases such as Crohn's disease or ulcerative colitis. Flavonoid phytoalexins bind to receptors in the cells and have antioxidant properties that help protect cells from damage caused by free radicals in the body. This makes them useful for treating heart disease and cancer. Flavonoid phytoalexins can also treat neurodegenerative diseases such as Parkinson's disease and Alzheimer's disease because they slow down the progression of these diseases by preventing damage caused by free radicals on nerve cells in our brains (Bizuneh, 2021).

There are three different types of food phytoalexins: saponin, lectin, and proteinase inhibitors. The structure of each one is different, but they all share some common characteristics: they are water-soluble and heat-stable; they have antimicrobial activity; they can interact with host cells; and their presence can change the activity of enzymes in the body (e.g., digestive enzymes). The most commonly known source for these compounds is peppers—peppers contain more than 50 different types of bioactive phytoalexins! Chili peppers, for example, contain about 15 different types of bioactive compounds (called capsaicinoids), which may help you burn fat faster during exercise or make it easier to fall asleep at night by reducing stress levels during the day (and by keeping blood pressure under control).

Chemical classification of phytoalexins: polyphenols, terpenoids, and alkaloids

They can be classified into three types: polyphenols, terpenoids, and alkaloids. The chemical classification of these compounds is a topic of interest for researchers and food scientists. The first group of compounds that can be classified as phytoalexins are polyphenols. Polyphenols are compounds that contain two or more phenolic hydroxyl groups. The classifications of phytoalexins come next in line: terpenoids, which contain one phenolic hydroxyl group; and alkaloids, which contain two or more phenolic hydroxyl groups.

The general chemical classification of food phytoalexins is based on their chemical structure: monomeric, oligomeric, or polymeric forms. The monomeric forms can be further grouped according to their hydrolysis products. The main monomeric compounds are anthocyanidins, which have high antioxidant activity; flavonoids and isoflavonoids, which have anti-inflammatory activities; and proanthocyanidins (PACs), which have anti-arteriosclerotic properties. Oligomeric compounds include procyanidin oligomers (such as cyanidin-3-O-glucoside) with potential antioxidant activity; *glucosylated anthocyanins* (glucosyl gallic acid); and flavanols (flavanols such as catechin). Polymeric compounds include PACs such as procyanidin dimers and trimmers.

Polyphenols

These compounds may also be referred to as phenolic compounds. Polyphenols are found in many types of plants, and they are often used as antioxidants. They are also known to have antibacterial, antiviral, and antifungal properties. In addition to their antioxidant properties, polyphenols have been shown to play a role in preventing cancer and other diseases. They are also a type of flavonoid that helps protect cells from damage caused by reactive oxygen species (ROS). These substances play an important role in metabolism, inflammation, and immune function. Flavonoids in polyphenols are also called tannins because they come from plants like oak trees.

Flavonoids can be divided into two groups: flavanones and flavan-3-ols (also known as catechins). These groups contain different numbers of OHs than other types of polyphenols do; therefore they contribute to the color and taste of fruits and vegetables when they're eaten by people or animals. For example the red color in apples is due to anthocyanidins that contain 3 OHs per molecule; but the yellow color in carrots comes from chlorogenic acid (Aluko, 2012).

Polyphenols can also be isolated from foods like onions and garlic. They may be called phenolic acids, tannins, flavonoids, or catechins. Some studies have shown that certain polyphenol compounds may help protect against cardiovascular diseases such as heart attack and stroke by reducing the risk of inflammation in the body (Bizuneh, 2021).

Terpenoids

Terpenoids are a class of chemicals that produce the flavors, fragrances, and colors found in foods. These compounds are derived from plants and are also known as phytoalexins. The three most common classes of terpenoid compounds include:

- Sesquiterpenes - these are terpenoid compounds that contain more than one isoprene unit
- Terpenoids - these are terpenoid compounds in which the first carbon atom is bonded to four double bonds at once
- Nonisoprenoids

They are produced by plants as a defense against herbivores, pathogens, and insects. While terpenoids can be found in many different plants, they are especially common in fruits that are high in vitamin C, such as citrus fruits. For example, broccoli contains betulinic acid, which acts as an anticarcinogen by inhibiting the growth of cancer cells. The isoprenoid class includes a variety of compounds with a similar structure but different chemical properties; examples include limonene from citrus fruits and linalool from lavender. Nonisoprenoid compounds include monoterpenes (such as carvone), sesquiterpenes (such as α-pinene), and triterpenes (such as β-caryophyllene).

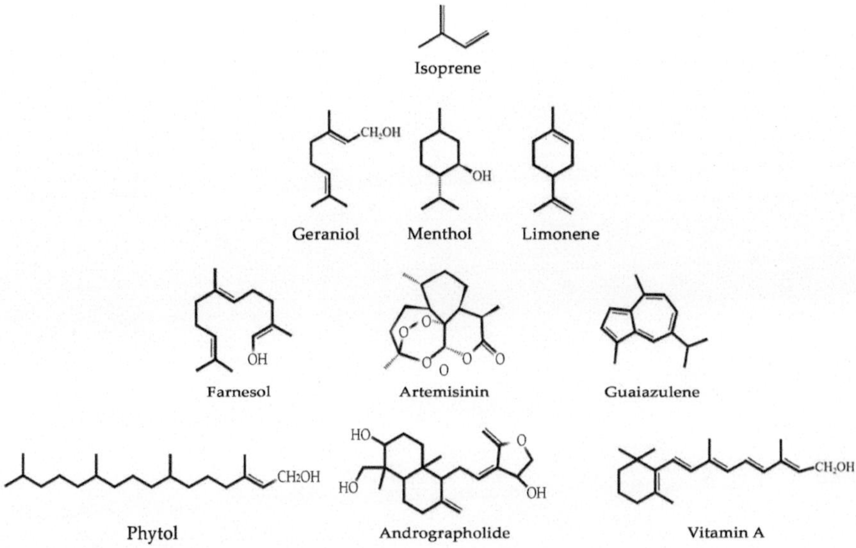

Figure 4.4. Examples of different classes of terpenoids.

Source: Fan M, Yuan S, Li L, Zheng J, Zhao D, Wang C, Wang H, Liu X, Liu J. Application of Terpenoid Compounds in Food and Pharmaceutical Products. *Fermentation.* 2023; 9(2):119. https://doi.org/10.3390/fermentation9020119

Alkaloids

Alkaloids are a chemical class that includes compounds with nitrogen as an essential functional group. They are naturally occurring compounds that can be extracted from plants. Alkaloids are found in many different types of food, including cereals, fruits, nuts, and vegetables. Alkaloids help protect plants from insects, deter herbivorous animals, and even repel predators. Alkaloid content varies among plant species and can vary depending on the part of the plant where it was produced. Alkaloids are found in all living organisms, where they are used as a tool for defense against pathogens, or to regulate cell growth and development. Alkaloids are found in plants and animals and are produced by plants to protect themselves from insect or fungal attacks or to ward off other organisms that might eat them. For example, the plant tobacco is poisonous to insects, but its alkaloids help it deter predators. The alkaloids that makeup tobacco also occur naturally in many other foods: mushrooms, potatoes, cauliflower, chocolate candy bar wrappers, parsley leaves, guava fruit, and cherries (Aluko, 2012).

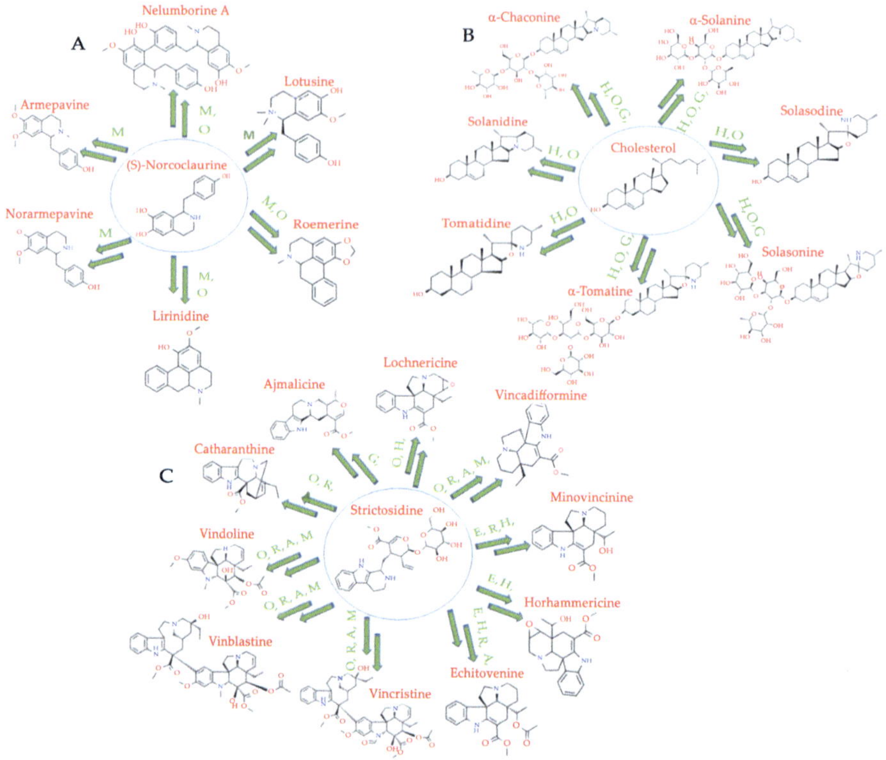

Figure 4.5. Skeletal structure of true alkaloids.

Source: Bhambhani S, Kondhare KR, Giri AP. Diversity in Chemical Structures and Biological Properties of Plant Alkaloids. *Molecules*. 2021; 26(11):3374. https://doi.org/10.3390/molecules26113374 \

Phytoalexins in cruciferous vegetables: glucosinolates and iso-thiocyanates

The phytoalexins are a group of compounds that are found in the Cruciferae family and are produced by plants to fight off pathogens. These compounds can be divided into two groups based on their chemical structure: glucosinolates, which include protease inhibitors (such as indole-3-carbinol); and isothiocyanates, which include phenolic compounds (such as sulforaphane).

Cruciferous vegetables are a group of plants that includes broccoli, cabbage, cauliflower, kale, and Brussels sprouts. These plants contain phytoalexins that are thought to help ward off diseases like cancer.

Glucosinolates are converted into isothiocyanates by enzymes called myrosinase and thioglucosidase. Phytoalexins in cruciferous vegetables are present in the plant's cell wall and include glucoraphanin (the precursor to glucosinolates) and sulforaphane (the precursor to isothiocyanates).

The process of making phytoalexins starts with the ingestion of cruciferous vegetables by plants. This causes the production of enzymes that break down these compounds into their phytochemicals, which can then be absorbed by humans after eating those vegetables. The quantities of these phytochemicals within the crucifers vary depending on their age, so older or more mature plants will have more than younger or less mature ones do. Glucosinolates are the most common type of phytoalexin found in cruciferous vegetables.

They are formed from the breakdown of amino acids in the cell walls of cabbage. Glucosinolates are also referred to as isothiocyanates because they can react with thiol groups (–SH) on proteins to form isothiocyanates (ITC).

There are more than 100 glucosinolates in multiple species of cruciferous vegetables and their glucosinolate content varies by strain. The most well-known isothiocyanate is sulforaphane (SFN), which has been studied extensively for its anticarcinogenic effects. Sulforaphane has also been shown to protect against oxidative damage caused by free radicals or reactive oxygen species (ROS) (Bizuneh, 2021).

Phytoalexins in berries: anthocyanins and ellagitannins

Phytoalexins are chemicals introduced into food as a result of the interaction between plants and their environment. The most common types of phytoalexins in foods include anthocyanins, ellagitannins, and isoflavones. Anthocyanins and ellagitannins are found in many different types of fruits, while isoflavones are found in soybeans and soy products.

Anthocyanins and ellagitannins are two phytoalexins that are present in berries. Anthocyanins are pigments found in many types of berries, including blueberries, blackberries, and red raspberries. Ellagitannins are present in grapefruit peel and have been shown to have antioxidant properties. Anthocyanins and ellagitannins are often known as "flavonoids", which is an umbrella term for a group of naturally occurring chemicals that impart vibrant colors to plants (and also play important roles in human health).

Anthocyanins and ellagitannins help to protect berries from pathogens by adding color to the fruit and also by providing a barrier that prevents bacteria from entering the berries (Aluko, 2012).

Phytoalexins in herbs and spices: phenolic acids and essential oils

Herbs and spices are known to contain a variety of phytoalexins, including phenolic acids and essential oils. Phenolic acids are a type of phytoalexin found in many herbs and spices, such as thyme, oregano, and rosemary. They have been shown to have antimicrobial, antioxidant, and anti-inflammatory properties. These compounds work by disrupting the cell membranes of bacteria and fungi, preventing their growth and replication. Essential oils are another type of phytoalexin found in herbs and spices. They are volatile compounds that give plants their characteristic aroma and flavor. Essential oils have been used for centuries for their medicinal properties, including their antimicrobial, antiviral, and anti-inflammatory effects. Some examples of essential oils found in herbs and spices include eucalyptol in eucalyptus, thymol in thyme, and carvacrol in oregano.

Sources of phytoalexins: plant parts and accumulation patterns

The sources of phytoalexins can vary depending on the plant species, but they are generally produced in plant parts that are exposed to stress. One of the most common sources of phytoalexins is the plant's leaves. Leaves are often the first line of defense against pathogens and predators, and they produce phytoalexins in response to stress. For example, tea leaves produce catechins in response to insect damage, which have been shown to have antimicrobial and antioxidant properties. Another source of phytoalexins is the plant's roots. Roots are often exposed to stress from soil-borne pathogens and produce phytoalexins to protect themselves. For example, ginger roots produce gingerols and shogaols in response to soil-borne pathogens, which have been shown to have anti-inflammatory and antioxidant properties.

Plant stems and bark are also sources of phytoalexins. These plant parts are often exposed to stress from physical damage or pathogens, and they produce phytoalexins to protect themselves. For example, grapevine stems produce resveratrol in response to fungal infection, which has been shown to have anti-inflammatory and antioxidant properties. The accumulation patterns of phytoalexins can also vary depending on the plant species and the type of stress. Some plants produce phytoalexins continuously, while others only produce them in response to stress. Additionally, the amount and type of phytoalexins produced can vary depending on the severity and duration of the stress.

Factors affecting phytoalexin content: genetic and environmental influences

Phytoalexin content is influenced by genetic and environmental factors. The genetic factors that affect the phytoalexin content include the genes coding for enzymes that break down toxins, such as phenylalanine ammonia-lyase (PAL) and tyrosine ammonia-lyase (TAL). The presence of these genes is associated with lower levels of phytoalexin content in certain plants. Environmental factors that can influence phytoalexin content include soil type, type of fertilizer used, and rainfall. The amount of rainfall a plant receives will determine how much phytoalexin it produces (Bizuneh, 2021).

Genetic makeup determines the amount of phytoalexin the plant produces in response to a given stress or disease. This is because certain genes are activated when a plant experiences stress or disease, causing them to produce more phytoalexin than normal. These genes are referred to as "stress genes" because they activate when a plant experiences stressful conditions like drought or cold temperatures. Environmental conditions can also affect phytoalexin content. For example, if you grow your plants indoors under fluorescent lights instead of natural sunlight, their rate of photosynthesis will be slower than outdoors where there is more sunlight available for photosynthesis purposes. This will result in lower levels of phytoalexin in your plants due to decreased photosynthetic activity compared with outdoor plants.

Plants that have a greater resistance to pathogens will therefore have higher levels of phytoalexin content and therefore be more effective in protecting themselves from microbial attack. Genetic factors can affect phytoalexin content in two ways. First, they can affect the amount of phytoalexin in a plant. For example, some plants produce high amounts of phytoalexin while others do not. It is believed that this variation is due to genetic differences between plants and their ability to produce phytoalexin. Second, genetic factors also determine how well a plant will perform concerning phytoalexin content. For example, some plants grow very quickly and therefore produce higher levels of phytoalexin than other plants. If you want your plant to grow faster than it might be worth growing a species with a high level of phytoalexin content because it will produce more in less time without sacrificing quality or taste. Many environmental factors can affect phytoalexin content, including temperature and humidity. These factors affect the rate at which phytoalexins are produced by plants. Higher temperatures and lower humidity generally increase the rate at which phytoalexins are produced, as does increase stress on plants from insects or other pests (Aluko, 2012).

Extraction and analytical techniques for studying phytoalexins

Extraction is a process by which the phytoalexins in a food item are extracted and separated from other compounds. The extraction process is one of the most important steps in studying phytoalexins because it allows researchers to isolate and identify the chemical compounds present in a sample. Once extracted, these compounds can then be analyzed using various methods such as GC/MS, HPLC/UV or LC/MS/MS. Different extraction methods have been developed to separate phytoalexin compounds from plant parts such as leaves or stems. The most common methods include Soxhlet extractions which involve heating plant material with water at high temperatures (up to 100°C) until all soluble matter has been extracted into the liquid phase (Bizuneh, 2021).

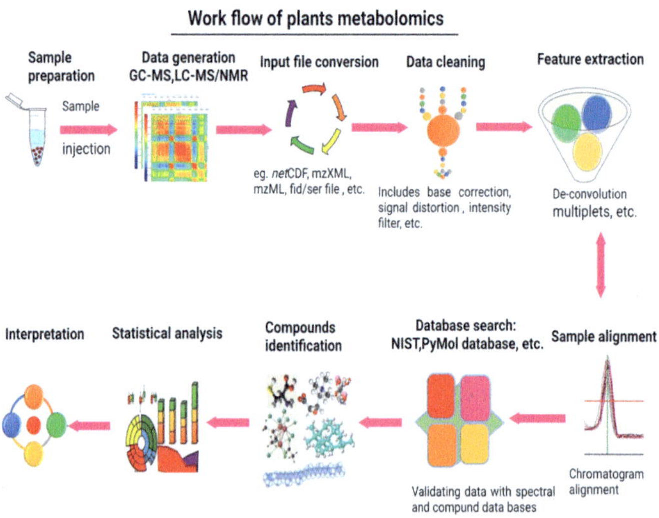

Figure 4.6. Plant Metabolomics: An Overview of the Role of Primary and Secondary Metabolites against Different Environmental Stress Factors

Source: Salam U, Ullah S, Tang Z-H, Elateeq AA, Khan Y, Khan J, Khan A, Ali S. Plant Metabolomics: An Overview of the Role of Primary and Secondary Metabolites against Different Environmental Stress Factors. *Life*. 2023; 13(3):706. https://doi.org/10.3390/life13030706

To extract phytoalexins from foods, one must first be aware of the different types of phytoalexins. There are two main groups of phytoalexins:

cereal-derived and noncereal-derived. Cereal-derived phytoalexins include isothiocyanates, indoles, and polyphenols (including tannins). Noncereal-derived phytoalexins include curcuminoids and flavonoids. The extraction technique used will depend on which type of phytoalexin is being studied. The most common extraction technique for noncereal-based phytoalexins involves heating the plant material with an organic solvent such as methanol or ethanol to extract the active ingredient into the solution. This process can take several days to complete depending upon how much solvent is used initially; however, it does not require special equipment other than a stovetop or microwave oven. The extracted solution may then be filtered through filter paper to remove any remaining solids or sludge left over from the extraction process itself (such as seeds or stems.).

4.4 BIOACTIVITY OF FOOD PHYTOALEXINS: ANTIMICROBIAL PROPERTIES

Food phytoalexins are antimicrobial compounds found in foods. They are produced by plants as a natural response to pathogens that invade the plant cell. Phytoalexins have been demonstrated to have antimicrobial properties, including the ability to inhibit the growth of bacteria and fungi. Some phytoalexins have also been shown to be effective against viruses. Food phytoalexins can be found in a variety of plants, but they are most abundant in cruciferous vegetables such as broccoli, cabbage, cauliflower, collard greens and kale. These compounds can be produced by plants to defend themselves against pathogens or insects. They also have antimicrobial properties, which means they can fight bacteria, fungi, and other microorganisms (Bizuneh, 2021).

There are many types of phytoalexins, but a few are well-known for their ability to fight off harmful organisms. The most well-known phytoalexin is apigenin, which has been used in traditional medicine for thousands of years. It has been used in many different forms such as tea and supplements to help lower cholesterol levels in the body. Another popular phytoalexin is luteolin, which is used as an anti-inflammatory agent and has been shown to decrease inflammation in patients with rheumatoid arthritis (RA). These compounds can have antimicrobial effects against a variety of different species of bacteria, including some that cause foodborne illness. Studies show that some phytoalexins can inhibit the growth of bacteria at concentrations lower than those needed to inhibit the growth of other microorganisms, like plant pathogens. These compounds may also be used as preservatives because they have antimicrobial effects without negatively affecting food quality or taste (Aluko, 2012).

Antimicrobial properties are widely known to be beneficial in the prevention of human disease. For example, garlic is used to treat colds, flu, and other illnesses because it has antimicrobial properties that help prevent the growth of certain bacteria in the body. Garlic also reduces the risk of cardiovascular disease by reducing blood pressure and lowering cholesterol levels.

Mechanisms of antimicrobial activity of phytoalexins

The mechanisms of antimicrobial activity of phytoalexins include:

- Inhibition of cell wall synthesis: Phytoalexins can inhibit the synthesis of bacterial cell walls by disrupting the formation of peptidoglycan and other essential components.
- Disruption of membrane integrity: Phytoalexins can damage the bacterial cell membrane by inserting themselves into the lipid bilayer, causing leakage of cellular contents and cell death.
- Inhibition of protein synthesis: Phytoalexins can interfere with bacterial protein synthesis by binding to ribosomes and other essential enzymes, thereby preventing the synthesis of new proteins.
- Oxidative stress: Phytoalexins can induce oxidative stress in bacteria by generating reactive oxygen species (ROS) that damage cellular components and disrupt metabolic processes.

Synergistic interactions of phytoalexins with conventional anti-microbial agents

The use of phytoalexins as standalone antimicrobial agents is limited due to their low potency and narrow spectrum of activity. This has led to an interest in exploring the potential of combining phytoalexins with conventional antimicrobial agents to enhance their efficacy. Such combinations have been shown to exhibit synergistic interactions, where the combined effect is greater than the sum of their individual effects. There are several mechanisms by which phytoalexins can enhance the activity of conventional antimicrobial agents. These include:

- Potentiation of cell wall damage: Phytoalexins can enhance the activity of antibiotics that target the bacterial cell wall, such as β-lactams, by disrupting the integrity of the cell wall and facilitating the entry of the antibiotic into the cell.

- Inhibition of efflux pumps: Efflux pumps are a common mechanism of resistance in bacteria, where they actively pump out antibiotics from the cell. Phytoalexins can inhibit the activity of efflux pumps, thereby increasing the intracellular concentration of antibiotics and enhancing their activity (Bizuneh, 2021).
- Disruption of biofilms: Biofilms are communities of microorganisms that are embedded in a protective matrix and are often resistant to antimicrobial agents. Phytoalexins can disrupt biofilms by inhibiting the production of extracellular polymeric substances that make up the matrix, thereby increasing the susceptibility of the microorganisms to antimicrobial agents.
- Modulation of host immune response: Phytoalexins can modulate the host immune response by enhancing the production of cytokines and other immune mediators that can help to clear infections. This can enhance the activity of antibiotics by complementing their direct antimicrobial activity.

Several studies have demonstrated the synergistic interactions between phytoalexins and conventional antimicrobial agents. For example, the combination of resveratrol, a phytoalexin found in grapes, with ampicillin or gentamicin was shown to enhance the activity of these antibiotics against methicillin-resistant Staphylococcus aureus (MRSA) and *Pseudomonas aeruginosa*, respectively. Similarly, the combination of berberine, a phytoalexin found in several plants, with ciprofloxacin was shown to enhance the activity of this antibiotic against Escherichia coli (Aluko, 2012).

The use of phytoalexins in combination with conventional antimicrobial agents has several advantages. First, it can enhance the efficacy of existing antimicrobial agents, thereby reducing the risk of resistance development. Second, it can broaden the spectrum of activity of existing antimicrobial agents, thereby increasing their usefulness in the treatment of infections caused by multidrug-resistant microorganisms. Thirdly, it can reduce the dose of conventional antimicrobial agents required to achieve a therapeutic effect, thereby reducing the risk of toxicity and side effects.

Phytoalexins and antibiotic resistance: potential alternatives to combat resistance

Phytoalexins are secondary metabolites produced by plants in response to pathogen attacks. These compounds are known to have antimicrobial properties and have been extensively studied in recent years. Phytoalexins are produced by a wide range of plants, including fruits, vegetables, and grains. Some of the most well-known phytoalexins include resveratrol, found in grapes, and brassinin, found in broccoli. One of the main advantages of phytoalexins is that they are natural compounds and are therefore less likely to cause side effects compared to synthetic antibiotics. Additionally, phytoalexins have broad-spectrum activity, meaning they can target a wide range of pathogens. This is in contrast to antibiotics, which are often specific to certain types of bacteria.

Another potential advantage of phytoalexins is that they may be less likely to cause resistance compared to antibiotics. This is because phytoalexins are complex compounds that are difficult for bacteria to adapt to. In contrast, antibiotics often target specific mechanisms within bacteria, making it easier for bacteria to develop resistance. Antibiotic resistance is a major problem in modern medicine, and there is a growing need for alternative treatments. Antibiotic resistance occurs when bacteria evolve to resist the effects of antibiotics, making it difficult to treat infections. This is a serious problem, as it can lead to longer hospital stays, higher healthcare costs, and even death. One potential solution to antibiotic resistance is the use of alternative treatments, such as phytoalexins. Another potential solution is the development of new antibiotics that are less likely to cause resistance. However, this is a difficult task, as bacteria are constantly evolving and adapting to new antibiotics.

Antioxidant activity of food phytoalexins and health benefits

Food phytoalexins are antioxidant compounds in fruits and vegetables. They are known for their anti-inflammatory, anti-atherosclerotic, and anti-cancer properties. There is evidence that phytoalexins have therapeutic effects on cardiac diseases, diabetes mellitus, neurodegenerative diseases, cancer and inflammatory bowel disease (IBD). There is increasing interest in the potential use of foods rich in phytoalexin content as preventive agents against disease. These compounds protect the plants from oxidative damage caused by reactive oxygen species (ROS). Antioxidants help to maintain healthy cells and tissues, but they can also protect against chronic diseases such as cancer and heart disease (Bizuneh, 2021).

There has been much interest in the health benefits of these compounds because they are known to have anti-inflammatory properties. This means that they may be able to reduce pain and inflammation associated with arthritis, heart disease, cancer, and other diseases. They may also be able to prevent damage from free radicals in your body by protecting cells from oxidative stressors like cigarette smoke or pollution. These antioxidants are found in many types of foods including fruits and vegetables like berries, apples, cranberries, cherries and grapes; herbs such as basil; spices like turmeric; sea foods like shrimp; whole grains such as brown rice; nuts like almonds and walnuts; legumes such as black beans.

4.5 PHYTOALEXINS AND CANCER PREVENTION: MECHANISMS AND POTENTIAL THERAPEUTIC APPLICATIONS

Phytoalexins have been studied for their potential use as cancer preventatives. They are known to inhibit tumor growth by causing apoptosis (programmed cell death) in cancer cells and by inducing inflammation that can kill cancer cells as well as normal cells. In addition, phytoalexin-based drugs have been shown to protect against colon carcinogenesis (cancerous growth). These chemicals have been found to have significant anticancer properties. In some cases, phytoalexins have been proven to be more effective than conventional cancer treatments.

Phytochemical research has revealed that many phytoalexins have potential applications in the prevention of cancer. Some of these compounds can kill cancer cells directly or indirectly by blocking the cell cycle or activating apoptosis (programmed cell death). The effectiveness of phytochemical treatment depends on their potency and concentration, as well as on how they are delivered to the target tissue (exogenous vs. endogenous). Some phytochemicals may be used as therapeutic agents for treating various types of cancers. One example is *epigallocatechin gallate* (EGCG). EGCG is a major component of green tea and has been shown in studies to inhibit proliferation, induce apoptosis, and inhibit angiogenesis (growth of new blood vessels) in tumors (Aluko, 2012).

Scientists are still learning how these chemicals work at a molecular level—but there's evidence that they may be involved in protecting against tumor formation by interfering with signaling pathways that promote cancer growth. The first major study of phytoalexins and cancer prevention was published in 2001 by researchers at the University Of California Davis School Of Veterinary Medicine (UC Davis). They found that adding fermented soybean or cabbage extracts to rodent diets reduced the incidence of tumors in mice by up to 50%. Other studies have shown that other phytoalexins have anti-cancer effects on human cells in culture. In one study published in 2015, scientists found that broccoli extract inhibited tumor growth in mice with lung metastases from breast cancer cells. A more recent study from 2016 found that broccoli extract inhibited the growth of lung cancer cells in rats with metastatic disease. In addition, research has shown that drinking green tea increases the concentration of a specific phytoalexin called ellagitannin in humans.

Phytoalexins and apoptosis (programmed cell death) in cancer cells

Apoptosis is a normal process by which cells eliminate damaged or unnecessary cells. However, it can also be triggered inappropriately, causing cancer. When this happens, the cell stops dividing and begins to die. This is called apoptosis. Phytoalexins are substances produced by plants that have anti-cancer properties, including apoptosis-inducing activity. They have been shown to inhibit tumor growth in animal studies and may also help prevent cancer in humans. During apoptosis, the cell changes its DNA and proteins to prepare it for cell suicide. These changes include condensation of chromatin, formation of apoptotic bodies, and release of cytoplasmic substances. Apoptosis plays an important role in cancer development, as it is a way for cells that have been damaged or mutated to undergo apoptosis and become parts of the tumor mass (Bizuneh, 2021).

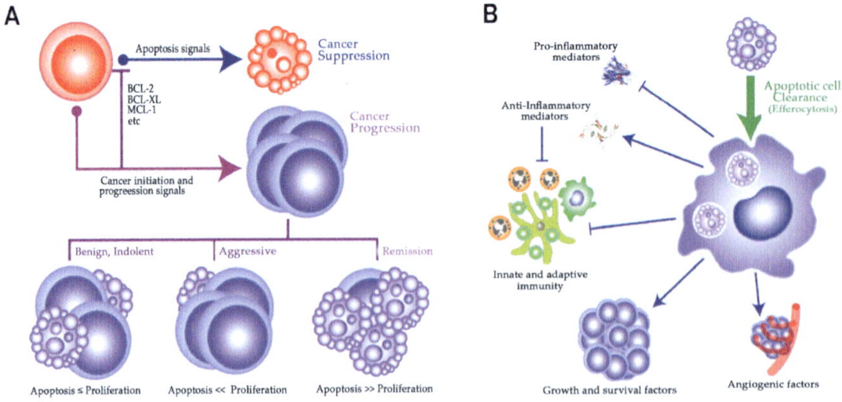

Figure 4.7. Apoptosis (Programmed cell death) Cell Biology.

Source: Morana, O., Wood, W., & Gregory, C. D. (2022). The Apoptosis Paradox in Cancer. International Journal of Molecular Sciences, 23(3), 1328. MDPI AG. Retrieved from http://dx.doi.org/10.3390/ijms23031328

Phytoalexins are compounds produced by plants that can produce cytotoxic activity against cancer cells. Phytoalexins may exert their effects through existing pathways that regulate apoptosis or through new pathways that are not yet known. The most common phytoalexin found in plants is allicin, which has been shown to trigger apoptosis in human pancreatic cancer cells. In apoptosis, the cell undergoes changes that result in its break-up into small pieces, or apoptotic bodies. Apoptotic bodies themselves can be seen as part of the normal process of removing unwanted cells from the body; however, if they accumulate in large numbers, they can become a problem for the organism. The mechanisms that cause phytoalexin production are not fully understood, but research has shown that these compounds modulate protein expression levels in response to stress conditions such as heat shock or cold shock (this effect is called upregulation) (Aluko, 2012).

Phytoalexins as modulators of oxidative stress and DNA damage

Oxidative stress occurs when there is an imbalance between the production of ROS and the ability of the body to detoxify them. ROS are highly reactive molecules that can cause damage to DNA, proteins, and lipids, leading to cellular dysfunction and disease. Phytoalexins have been shown to have antioxidant properties, meaning they can neutralize ROS and prevent oxidative damage. Several studies have shown that phytoalexins can modulate oxidative stress by increasing the activity of antioxidant enzymes such as superoxide dismutase (SOD) and catalase. For example, resveratrol, a well-known phytoalexin found in grapes, has been shown to increase SOD activity and reduce ROS levels in cells. In addition to their antioxidant properties, phytoalexins have also been shown to modulate DNA damage. DNA damage can occur from a variety of sources, including exposure to UV radiation, environmental toxins, and oxidative stress (Bizuneh, G. K. 2021).

Phytoalexins have been shown to protect against DNA damage by increasing the activity of DNA repair enzymes such as poly(ADP-ribose) polymerase (PARP). One study showed that resveratrol can protect against DNA damage caused by UV radiation by increasing PARP activity and reducing the formation of DNA adducts.

Another study showed that brassinin, a phytoalexin found in broccoli, can protect against DNA damage caused by environmental toxins by increasing the activity of DNA repair enzymes. Overall, the ability of phytoalexins to modulate oxidative stress and DNA damage makes them promising candidates for the prevention and treatment of diseases associated with these processes. For example, phytoalexins may be useful in the prevention of cancer, as oxidative stress and DNA damage are known to be major contributors to the development of cancer.

4.6 FACTORS AFFECTING PHYTOALEXIN CONTENT IN FOODS

Many factors affect the phytoalexin content in foods. The genetic makeup of the plant itself is one of the most important factors in determining how much phytoalexin a plant produces. Plants with different genetic backgrounds will have varying levels of phytoalexin production, which means that there may be differences in overall phytoalexin content among different plants. Processing methods can also influence phytoalexin content in foods: for example, when cooking or freezing food at high temperatures, the process breaks down some of the protective chemicals in the plant's cells. This change can then impact levels of active compounds that would otherwise be included in these foods (Aluko, 2012).

The genetically-determined level of phytoalexin content in a plant is inherited from one generation to the next. This is due to a mutation in the gene that codes for phytoalexin biosynthesis. The mutation can occur in both parents or one parent. The mutation causes an increase in the amount of phytoalexin present at an early stage of life, which results in higher levels of phytoalexin being produced at later stages of life. A high level of phytoalexin production leads to higher levels of protection against pathogens and insects. However, this also means that there will be more phytoalexins produced when the plant is subjected to stressors such as drought or frost as these stressors may cause mutations that lead to increased levels of phytoalexin production and thus higher levels of protection against pathogens and insects; however, it can also lead to lower growth rates and yield due to increased susceptibility to these stresses. For example, if you have a plant that has a high level of phytoalexin, you will likely produce plants with a higher level of phytoalexin as well. However, if you cross-breed these plants with another with a lower level of phytoalexin, you would not see any significant difference in your final crop's phytoalexin content because it is determined by the genetic makeup of the plants themselves.

The environmental conditions under which a plant grows can also affect its phytoalexin content. If a plant is exposed to higher levels of light or heat than others, it may produce more phytoalexin than other plants with similar genetic backgrounds. Other factors such as altitude and soil type can also impact phytoalexin production. For example, if there are more bugs or fungi around, they can eat up all the phytoalexin in plants and make less of it available to us. However, environmental factors don't always have an impact on phytoalexin content. For example, if you grow your food at home without any chemicals or pesticides or herbicides, then those chemicals won't be able to leach into your food when you eat it—which means that even though there might be fewer phytoalexins from chemicals used in your home garden than from chemical use on farms elsewhere. The climate can also influence phytoalexin content in plants. For example, high temperatures may increase phytoalexin production because heat affects enzymes that are involved with phytoalexin synthesis. However, if there is too much heat or not enough water during growth periods then this could lead to reduced levels of phytoalexin production due to changes in enzymes needed for its synthesis (Bizuneh, 2021).

4.7 FOOD PROCESSING TECHNIQUES AND PHYTOALEXIN STABILITY

Food processing techniques are methods used in processing food to remove undesirable components, such as the phytoalexins found in plants. One common technique is drying out the food item. Phytoalexins are most commonly found in foods that have been dried out for a long period or frozen. Drying removes water from the food item, which prevents them from forming ice crystals and damaging their structure. This is why dried fruits or vegetables often contain less phytoalexin content than fresh ones. Another common method used to preserve food items is canning or freezing. By removing moisture from these foods, they become less susceptible to damage from moisture loss during storage conditions such as humidity levels or temperatures. This reduces the amount of phytoalexin loss through spoilage due to heat exposure during processing or during storage conditions such as temperature fluctuations between hot and cold environments (i.e., refrigeration cycles) (Aluko, 2012).

Phytoalexin stability is an important aspect of food processing techniques. In fact, it can have such a significant effect on the quality of your finished product that it's difficult to overstate how important it is. The phytoalexin content in raw fruits and vegetables varies greatly depending on the species of plant and its stage of ripeness, but most plants contain between 0.5% and 3% phytoalexin. This compound is found in a wide range of foods, including grains, nuts, fruits, and vegetables. It has been found in some plant-based medications as well as animal products like milk. In order for phytoalexin to be effectively removed from a food product during processing or packaging, it must be broken down into smaller molecules that can be more easily washed away from surfaces that come into contact with your finished product. For example: if you're processing grain products like rice or wheat, you'll want to use techniques that break down the larger molecules of phytoalexin into smaller ones that can be rinsed off without leaving residue behind. There are two main ways to do this: physical methods and chemical methods (which include enzymatic treatment) (Bizuneh, 2021).

Chapter

5

BIOACTIVITY OF FLAXSEED LIGNANS

CONTENTS

5.1 INTRODUCTION

Flaxseed lignans are a naturally occurring source of health-promoting, biologically active compounds. The positive effects of phenolic compounds are primarily attributable to their antioxidant properties, which shield the epidermis from oxidative stress and stimulate collagen synthesis. FlaxseSed is also known as linseed or *Linum usitatissimum*. Flax is extensively cultivated throughout the world, with the majority of it coming from Canada. Fruit receptacles containing tiny brown seeds and delicate blue flowers are produced by flax plants. Since it is one of the most abundant plant sources of n-3 PUFA ALA, flaxseeds can be processed industrially to produce flax oil. 32-45% of the seed's mass consists of oil, of which 55-57% is alpha-linolenic acid (ALA), a type of n-6 PUFA, and 15-18% is linolenic acid. After oil extraction, 55-68 percent of the flaxseed mass remains as defatted flax flour. Typically, the oil is fed to animals. Flax oil is also utilized in paints, varnishes, and inks due to its rapid polymerization properties. On the other hand, flaxseed is frequently used as a functional food because it contains the maximum concentration of essential fatty acids derived from plants. In addition to ALA, flaxseed contains lignans, cyclic peptides, cyanogenic glycosides (CGs), flax proteins, soluble and insoluble fiber, and additional bioactive compounds. Whole flaxseed, powdered flaxseed, flaxseed oil, and partially defatted flaxseed meal are the four most common types of flaxseed available for human consumption (Balducchi, 2023). The flaxseed is composed of an embryo, two cotyledons, and a glossy brown shell. Lignans from flaxseed are frequently discussed in the scientific literature. In the meantime, the mechanisms underlying their functional and therapeutic properties are unknown. It is also essential to discover lignan synthesis and purification techniques that are scalable, inexpensive, and user-friendly. These systems may be able to meet the requirements of the pharmaceutical, cosmetic, and nutraceutical industries while producing market-competitive products (Ranveer, 2015).

Antihypertensive, antiatherogenic, cholesterol-lowering, anti-inflammatory, and arrhythmia-inhibiting properties have been associated with flaxseed consumption. Its high fiber content, omega-3 fatty acid linolenic acid content, and antioxidant lignan, content have all been linked to these cardiovascular health benefits. Proteins, cyclolinopeptides, and cyanogenic glycosides are additional potentially bioactive compounds present in flaxseed, but they are less well-known. These chemicals may be responsible for the cardiovascular benefits of flaxseed. This article will review the bioactive components of flaxseed in terms of their properties, biological effects, and proposed mechanisms of action, as well as the cardiovascular benefits of flaxseed supplementation. In addition, it will investigate enticing prospective research directions to uncover additional health benefits of flaxseed consumption.

5.2 HEALTH BENEFITS OF FLAXSEED

This plant has antioxidants, healthy fats, and fiber. Flaxseed currently comes as seeds, oil, and granules. Flaxseed supplements prevent constipation, diabetes, high cholesterol, heart disease, cancer, and other disorders. Flaxseed includes ALA, lignans, antioxidants, fiber, and protein. Nutrients may lower illness risk. Flaxseed in the diet is useful for treating many medical disorders, according to extensive and growing research. In t he late 20th century, limited knowledge of health benefits and ideal consumption techniques differ greatly from current information. Flaxseed benefits research has risen considerably. Research investigates flaxseed's nutritional and disease-fighting properties. The bioactive components of flaxseed that provide these health advantages and the forms needed for absorption have also been discovered. Animal and human studies on flaxseed diets and health consequences are few. Several trials were undertaken to employ the strictest preventative measures to safeguard the public, especially those with impaired immune systems, from dangerous illnesses. The food sector has added additional flaxseed-containing products due to the above discoveries, offering customers more alternatives for adding them to their diets. Due to its expanding marketability and public knowledge of its potential health advantages when added to a regular diet, flaxseed is expected to become economically significant for farmers, food processors, and retailers (Balducchi, 2023).

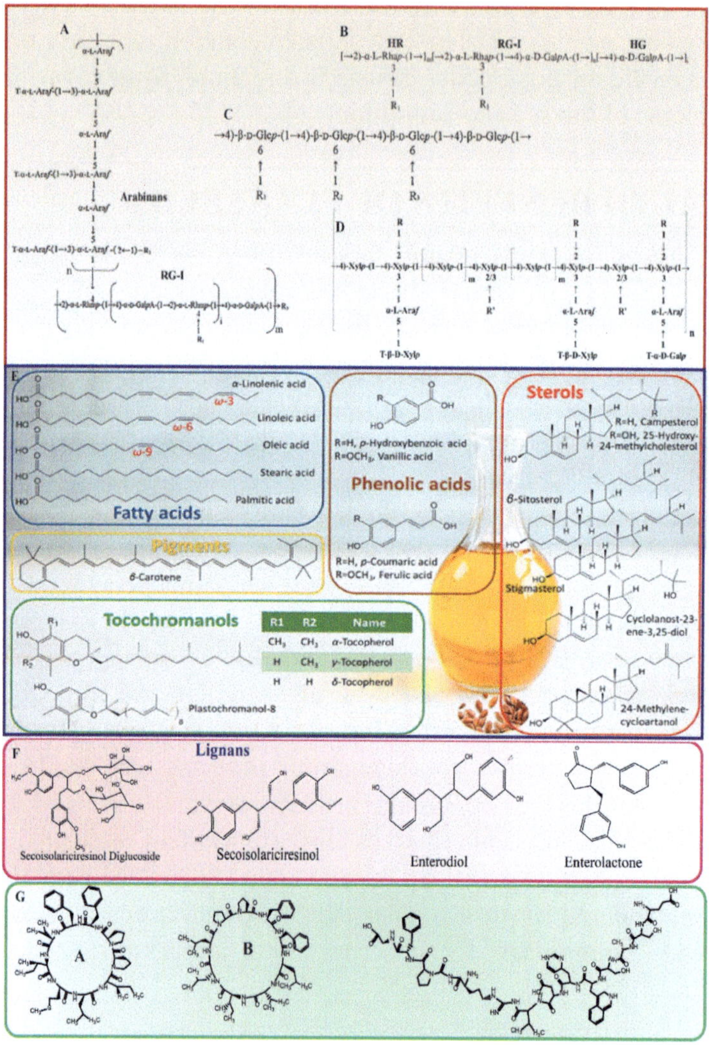

Figure 5.1. Nutritional profile of flaxseed.

Source: K Mueed A, Shibli S, Korma SA, Madjirebaye P, Esatbeyoglu T, Deng Z. Flaxseed Bioactive Compounds: Chemical Composition, Functional Properties, Food Applications and Health Benefits-Related Gut Microbes. *Foods*. 2022; 11(20):3307. https://doi.org/10.3390/foods11203307

Environmental variables, especially nutrition, affect health and illness. Dietary changes and pharmaceutical treatments are also showing promise for curing illnesses. This section summarizes flaxseed's health and medical breakthroughs.

5.2.1 Lowers cholesterol and treats hyperlipidemia

Hyperlipidemia is characterized by elevated total cholesterol, low-density lipoprotein cholesterol (LDL-C), triglyceride levels in the blood, or all three. Hyperlipidemia can be caused by environmental factors, inherited factors, or a combination of the two. A preponderance of patients with hyperlipidemia are acquired. As a risk factor for cardiovascular diseases such as coronary artery disease, stroke, and peripheral vascular disease, hyperlipidemia is second only to hypertension.

The primary causes of heart disease and stroke are hypertension and hypercholesterolemia It has been demonstrated that flaxseed and its constituents inhibit, reverse, and slow down hypercholesterolemic arthrosis. In hypercholesterolemic patients, hypercholesterolemic atherosclerosis may be the cause of coronary artery disease, peripheral arterial disease, and stroke. Statins, bile acid sequestrants, nicotinic acid, fibric acid, and evolocumab have all been used to reduce cholesterol levels in the blood. Despite a paucity of comparative studies, it appears that alternative nutraceuticals, such as flaxseeds and their constituents, can reduce blood lipids (Balducchi, 2023).Regular consumption of flaxseed increases blood flow and decreases triglyceride and cholesterol levels.

The dietary fiber in flaxseed can reduce cholesterol levels naturally by increasing lipid excretion in the intestines soluble flaxseed fiber can also entrain bile, thereby decreasing lipid reabsorption and enhancing excretion. Hyperlipidemia, or elevated levels of blood lipids such as triglycerides and cholesterol, is a major cardiovascular disease risk factor. Typical hyperlipidemia treatments include both medication and lifestyle modifications. Just recently, flaxseed was authorized as a food that reduces blood lipid levels. A variety of lignan secoisolariciresinol diglucoside (SDG) metabolites reduce blood lipid levels. SDG is present as an ester-linked copolymer with 3-hydroxy-3-methylglutaric acid (HMGA) in the flaxseed shell.

Nevertheless, purification methods such as copolymer hydrolysis and SDG enrichment are typically expensive. SDG polymer, which is a naturally occurring copolymer of SDG and HMGA, is a source of bioactive compounds that can be used to prevent hypercholesterolemia. Following the consumption of lignan copolymer, the stomach, and small intestines produce SDG and HMGA. The bioactive mammalian lignans secoisolariciresinol, enterolactone, and enterodiol are produced during the metabolism of SDG. Once distributed throughout the body, these metabolites begin to accumulate in the liver, kidney, epidermis, and other tissues and organs. These metabolites sequentially reduce blood lipid concentrations of cholesterol, triglycerides, low-density lipoprotein cholesterol, and lipid peroxidation products (Ranveer, 2015).

5.2.2 Rich in fiber

Flaxseed is a rich source of alpha linolenic acid, lignan secoisolariciresinol diglucoside, and fiber. These compounds exhibit bioactivity that is advantageous for the health of both humans and animals, owing to their anti-inflammatory, anti-oxidative, and lipid-regulating propertiesThere are several research studies that provides evidence for the efficacy of flaxseed in addressing a range of health concerns. Contemporary understanding of the advantageous properties of flaxseed and the optimal methods of its consumption diverges significantly from the comprehension prevalent in the latter part of the 20th century, which was characterized by a dearth of information. The quantity of research conducted on the advantages associated with the consumption of flaxseed has experienced a significant surge.

The present understanding encompasses the potential therapeutic and prophylactic applications of flaxseed, the dietary advantages it offers, the bioactive constituents that are typically responsible for conferring these benefits, and the requisite forms of flaxseed for bioactive absorption by the body. There is a paucity of information regarding the impact of flaxseed consumption on health outcomes, spanning from animal experimentation to human clinical trials. Numerous trials were carried out with utmost control measures to safeguard the general populace, especially those with weakened immune systems, against the risk of contracting diseases. The aforementioned findings have served as a stimulus for the food sector to diversify its flaxseed-infused product line, thereby expanding the range of choices available to consumers who wish to include flaxseed in their regular dietary regimen. It is expected that flaxseed will gain economic importance for farmers, food processors, and retailers in the coming years due to its growing availability in the market and the increasing public awareness of its potential health benefits when incorporated into daily diets.

It is currently acknowledged that environmental factors, including dietary habits, can exert a substantial influence on the preservation of health and the development of diseases. Furthermore, it is increasingly apparent that alterations in dietary intake, alongside pharmacological interventions, can be employed as a means of managing medical conditions. The objective of this research is to present a concise overview of the noteworthy discoveries derived from the investigations that have contributed to the eminence of flaxseed in the fields of health and medicine (Ranveer, 2015).

The three main bioactive constituents of flaxseed are alpha-linolenic acid (ALA), lignans, and fiber. There exist four primary variants of flaxseed that are commonly consumed by humans, namely whole flaxseed, powdered flaxseed, flaxseed oil, and flaxseed meal. For example, Pizzey Ingredients Inc. has recently launched flax "milk" in Manitoba, Canada. Flax milk is a potential substitute for traditional dairy milk options, such as almond milk. It is produced by blending finely ground flaxseed with purified water and other minor constituents. Flax milk is considered a suitable substitute for dairy milk owing to its elevated levels of ALA and lack of lactose and cholesterol. Compared to almond milk, this alternative possesses a greater number of health advantages and is appropriate for individuals with soy, nut, or gluten sensitivities (Balducchi, 2023).

Despite the nutritional value of a meal, individuals are likely to reject it if they find its flavor, texture, appearance, color, or aroma unappealing. Several undesirable qualities of flaxseed may harm its flavor. The two primary drawbacks associated with flaxseed are its bitterness and the potential for its exceptionally elevated omega-3 fatty acid concentration to undergo oxidation and spoilage. ALA is a highly susceptible fatty acid to oxidation due to its high concentration. Negative taste evaluations may result from the development of unpleasant flavors and musty fragrances due to the process of oxidation and eventual rancidity. The compound known as secoisolariciresinol diglucoside (SDG), which is present in flaxseed, exhibits a noteworthy antioxidant property. Moreover, flaxseed possesses a pleasant nutty fragrance and flavor, rendering it a prime component for a diverse range of culinary preparations. Research indicates that flaxseed has been successfully integrated into various food items such as food bars, muffins, bagels, bread, buns, tea biscuits, cinnamon rolls, and pasta. The inclusion of flaxseed in a product has an impact on its taste. Before the baking process, a proportion of 2 to 8% of the overall constituents (measured by weight) of flaxseed was incorporated into entrees intended for human consumption. Prolonged ingestion of flaxseed up to 40-50 g per day has been reported in the literature. Studies have been carried out on the efficacy of prolonged food ingestion by patients for one year in clinical trials. The results indicate that the dropout rates for both the placebo and flax groups were approximately 20%. This finding illustrates that to ensure participant retention over a prolonged duration of a study, noncompliance cannot be attributed to the consumption of flaxseed per se. rather, it is imperative to offer a broad range of meals that incorporate flaxseed to ensure daily dietary diversity.

Among the food items with a high flaxseed content are baked foods. . According to a study, the concentration and quantity of ALA in baked pastries remained unaffected after undergoing a baking process of two hours at a temperature of 178°C. The incorporation of flavoring agents into the baked preparation serves to obscure any unfavorable taste attributes resulting from the bitterness of flaxseed or slight rancidity. Several flavors, including banana and almonds, banana chocolate chip, cinnamon apple raisin, cinnamon raisin, cinnamon spice, gingerbread raisin, cappuccino chocolate chip, sunflower sesame, and cranberry orange, have been employed in scholarly investigations. Flavorings and compounds derived from flaxseed are likely to be preferred by the general public. According to an annual survey conducted to determine consumers' typical meal preferences, bagels were favored over other options such as muffins, snack bars, cookies, crackers, pasta, and sun-dried tomato rolls. According to the research findings, the flavors that were most preferred by consumers were cinnamon-raisin, cranberry-orange, apple-spiced, and sunflower-sesame. The advancement in the production of flax "milk" in recent times has facilitated the consumption and preservation of flaxseed (Balducchi, 2023).

Each of these dietary matrices possesses the potential to modify the stability of flaxseed. The product's stability is influenced by various factors, including the seed refinement process, storage temperature, and duration, and the type of flaxseed utilized, such as milled, whole, or flax oil. The process of milling, grinding, or pulverizing flaxseed results in the disruption of its sturdy protective seed coat, thereby exposing alpha-linolenic acid (ALA) and SDG to oxidative degradation. However, it is imperative to employ this approach to render these bioactive molecules attainable. According to research, the absorption of ALA by the body is more efficient when it is consumed in the form of oil or powder. Incorporating ground flaxseed into baked goods has been suggested as a potential strategy to inhibit the deterioration of alpha-linolenic acid (ALA) and SDG. According to research, the preservation of ALA and SDG, particularly in flax oil, can be improved by reducing storage temperatures and shortening storage periods (Ranveer, 2015).

There has been a recent expression of concern regarding the potential of flaxseed to contain certain substances that could potentially diminish the bioavailability and bioaccessibility of crucial minerals. Flaxseed contains phytic acids, linatine, protease inhibitors, and cyanogenic glycosides. Nevertheless, no detrimental impacts have been associated with these constituents in human research. It is plausible that the concentrations of said constituents within flaxseed may not be of sufficient magnitude to elicit physiological reactions. Nevertheless, it is advisable to evaluate the stated apprehension. Plant bio-breeding and/or food processing techniques can be employed to decrease the levels of these constituents in flaxseed.

Flaxseed contains a considerable amount of mucilaginous gum, which is a type of gel-forming fiber that undergoes minimal digestion as it passes through the gastrointestinal tract. The seeds typically contain between 8-12% of this substance. This particular form of soluble fiber has been found to enhance feelings of fullness and reduce the rate of nutrient absorption. The presence of fiber in flaxseed may hinder the digestion and absorption of certain nutrients. Flaxseed, being a low-carbohydrate food, is rich in both soluble and insoluble fiber. It has the potential to augment the speed and quantity of colon transit. The process of expeditious transit of food through the colon, leading to the elimination of toxins, is commonly referred to as a colon cleanse. Whilst the efficacy of colon cleansing remains a topic of debate, the impact of dietary fiber on satiety and weight reduction is widely acknowledged. The consumption of three tablespoons (30 g) of whole flaxseed can provide men and women with 22% and 34% of their daily recommended fiber intake, respectively. A diet that is rich in flaxseed has been found to enhance and sustain soft stools and regular bowel movements, while also increasing the weight and size of fecal matter (Ranveer, 2015).

5.2.3 Digestive health support

One of the most commonly acknowledged benefits of flaxseed is its capacity to improve digestive health. The administration of ALA has been shown to confer protection to the mucosal lining of the gastrointestinal tract while also exhibiting anti-inflammatory properties. Studies have shown that flaxseed provides benefits to individuals suffering from Crohn's disease and other gastrointestinal disorders. The consumption of dietary fiber has been found to facilitate the growth of beneficial intestinal microorganisms, even among individuals with sufficient digestive capacity. According to popular belief, the ingestion of flaxseed fiber can assist in the obstruction of waste and detoxification of the gastrointestinal tract. Adequate nourishment and a suitable environment are essential for the growth and survival of beneficial gastrointestinal microorganisms. These microorganisms thrive on intricate carbohydrates, such as those present in flaxseed gum. Flaxseed is widely recognized as a natural remedy for constipation owing to its significant concentration of both soluble and insoluble fiber. In the capacity of a nutritional adjunct, it facilitates the maintenance of consistent gastrointestinal motility (Balducchi, 2023).

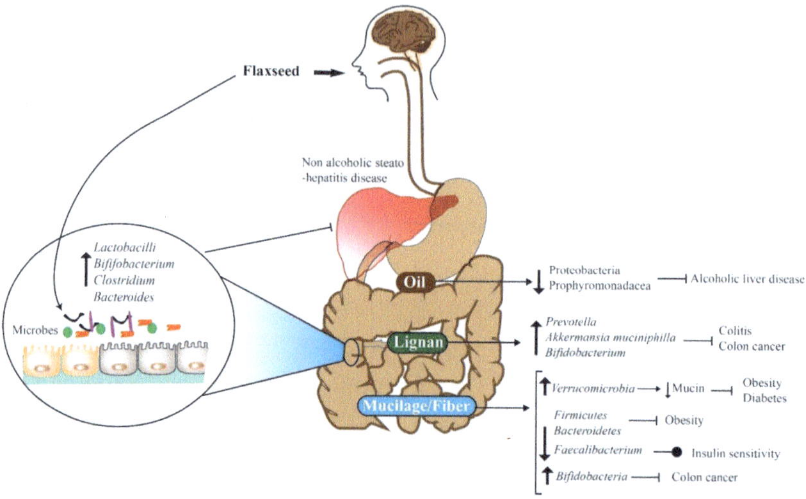

Figure 5.2. The effect of dietary flaxseed on gastrointestinal health.

Source: Mueed A, Shibli S, Korma SA, Madjirebaye P, Esatbeyoglu T, Deng Z. Flaxseed Bioactive Compounds: Chemical Composition, Functional Properties, Food Applications and Health Benefits-Related Gut Microbes. *Foods*. 2022; 11(20):3307. https://doi.org/10.3390/foods11203307

There is a growing interest within the scientific community regarding the impact of the microbiome on the health of both humans and animals. Currently, there is limited knowledge regarding the impact of flaxseed on the gut microbiota of both healthy and diseased populations. Before the commencement of this research, the prevailing notion was that the extensive lignans present in flaxseed necessitated the involvement of intestinal microorganisms for their digestion before they could permeate the human systemic circulation. According to research, milled flaxseed and defatted flaxseed meal are the most efficient in producing enterolignans, specifically enterodiol, and enterolactone, from SDG in comparison to other meals that contain SDG. The intestinal metabolism is driven by the biotransformational activity of specific intestinal bacteria, namely *Ruminococcus bromii* and *Ruminococcus lactaris*. Previous studies have indicated that the presence of *Lactobacillus acidophilus* and *Lactobacillus casei* is crucial for the process of flaxseed digestion and the attainment of enterodiol bioaccessibility.

Initial research has indicated that the consumption of flaxseed has the potential to modify the composition of the gastrointestinal microbiota in animals. The introduction of flaxseed into the dietary regimen of rodents has been observed to alter the quantity and diversity of Enterobacteriaceae present in their cecum and fecal matter, as reported in a previous study. The introduction of flaxseed in the diet leads to a significant modification in the colon's microbial composition. Specifically, there is a notable surge of 20 times in the prevalence of *Prevotella* spp. while the abundance of *Akkermansia muciniphila* experiences a substantial decline of 30 times. Previous studies have indicated that the incorporation of flaxseed and fish oil into a high-fat diet results in an elevation of Bifidobacterium levels within the gastrointestinal tract (Ranveer, 2015).

The manipulation of certain microbiome bacteria has the potential to impact the advancement of diseases. According to research, the consumption of flax oil may have a positive impact on alcoholic liver disease by decreasing the abundance of Proteobacteria and Porphyromonadaceae in the gastrointestinal microbiome. The suppression of aberrant crypt foci in the colon of rodents was observed through the increase in Bifidobacterium production by defatted flaxseed meal, which is a partially dilapidated variant of flaxseed, as reported in reference 81. The ingestion of flaxseed mucilage daily resulted in a modification of the prevalence of 33 distinct metagenomics species within the gastrointestinal microbiota (Balducchi, 2023).

An investigation into the potential of flaxseed as a substitute or mitigator of constipation medication usage among diverse at-risk populations, including the elderly, the infirm, and those residing in personal care facilities, would yield advantageous outcomes. Additionally, exploring the potential for flaxseed to reduce the adverse effects associated with conventional constipation medications would be of great value.

The outcomes of two randomized controlled trials investigating the efficacy of flaxseed in constipation treatment regimens have produced particularly fascinating findings. In the initial investigation, individuals with obstructed Type 2 diabetes consumed flaxseed that had been incorporated into baked biscuits. The consumption of flaxseed resulted in a decrease in constipation symptoms, weight, fasting plasma glucose, triglyceride, LDL, and HDL cholesterol levels. In a follow-up investigation concerning the identical patient cohort, it was observed that flaxseed enhanced all of the aforementioned traits and, notably, exhibited superior efficacy to psyllium in alleviating constipation symptoms. Further investigation is necessary to ascertain the potential of flaxseed in mitigating symptoms associated with irritable bowel syndrome (IBS). Flaxseed oil was found to be efficacious in the treatment of experimental diarrhea.

The considerable contribution of flaxseed and its high fiber content to the augmentation of fecal output in individuals with obstruction is noteworthy. Hence, it is unsurprising that flaxseed meals, despite being partially defatted, exhibited effectiveness as a laxative in individuals with normal bowel movements as well as those experiencing constipation. Research has demonstrated that flaxseed oil can regulate the movement of the intestines by interacting with the muscarinic receptors in the intestines and inhibiting the K+ channels.

5.2.4 Lower cancer risk

Furthermore, research suggests that the consumption of flaxseed may potentially decrease the likelihood of various types of cancer, such as breast, prostate, ovarian, and colon cancer. The ingestion of 30 grams of flaxseed daily could potentially decrease the likelihood of breast cancer development. Intestinal microorganisms can convert flaxseed lignans into mammalian lignans, namely enterolactone, and enterodiol. These chemicals exhibit biological effects that are similar to estrogen. The inclusion of phytoestrogens in the dietary intake of women may potentially mitigate the likelihood of developing breast cancer and other ailments that are linked to insufficient or erratic levels of estrogen. Flaxseed lignans have the potential to decrease the occurrence of endometriosis and ovarian cancer due to their similar mechanisms (Ranveer, 2015).

Research conducted on animals has demonstrated the efficacy of flaxseed in the treatment of diverse types of cancers. Breast cancer has been the primary focus in studies examining the impact of flaxseed on the human body. Both animal and human studies have demonstrated that the consumption of flaxseed significantly decreases the likelihood of developing breast cancer, as evidenced by references for animal studies and references for human studies. According to a meta-analysis of ten human studies, the consumption of flaxseed was associated with a reduction in tumor development among individuals diagnosed with breast cancer.

According to other research findings, flax has been associated with a decreased likelihood of mortality in individuals who have survived breast cancer, as well as providing a safeguard against the onset of early breast cancer. The nonsteroidal phytoestrogens known as flaxseed lignans exhibit molecular similarities to mammalian estrogens and have been observed to elicit comparable effects to estrogen in animal models. Before the detection of flax lignans in plasma, their degradation by intestinal microorganisms is a prerequisite. Most of the lignans that are consumed in the form of glucosides undergo deglycosylation by -glucosidases present in the gastrointestinal microbiota before being absorbed in the intestine. SDG undergoes hydrolysis to release secoisolariciresinol (SECO). SECO undergoes deglycosylation and demethylation processes, resulting in the formation of dihydroxy enterodiol. This compound is subsequently hydroxylated by microbiota, leading to the production of enterodiol. Enterolactone is the resultant compound obtained from the dehydration of enterodiol. Enterodiol and enterolactone undergo absorption and conversion into glucuronide and sulfate conjugates within the large intestine. The metabolic process for converting SDG to enterolactone is uniform across all genders, as indicated by the source. According to research on human pharmacokinetics, the emergence of enterolignans in the plasma occurs approximately 8 to 10 hours after the consumption of SDG. The average highest concentrations of enterodiol and enterolactone in the plasma are observed at 14.8 and 19.7 hours after consumption, respectively. According to a study, the standard elimination half-lives of enterodiol and enterolactone were found to be 4.4 and 12.6 hours, respectively. Empirical investigations in epidemiology have established a correlation between elevated concentrations of enterolactone in the bloodstream and a decreased occurrence of breast cancer, a reduced fatality rate, and improved longevity among women who have undergone menopause and have been diagnosed with breast cancer. According to clinical evidence, enterolactone has demonstrated efficacy in the prevention of various types of cancers such as breast, colon, prostate, gastrointestinal, and lung cancers (Balducchi, 2023).

Studies have demonstrated that the consumption of flaxseed or its constituents is associated with an elevated susceptibility to a diverse range of malignancies, beyond the scope of breast cancer. Studies have shown that the consumption of flaxseed has inhibitory effects on various types of malignancies, including prostate, lung, colon, cervix, endometrial, ovarian, and hepatocellular cancers (Ranveer, 2015).

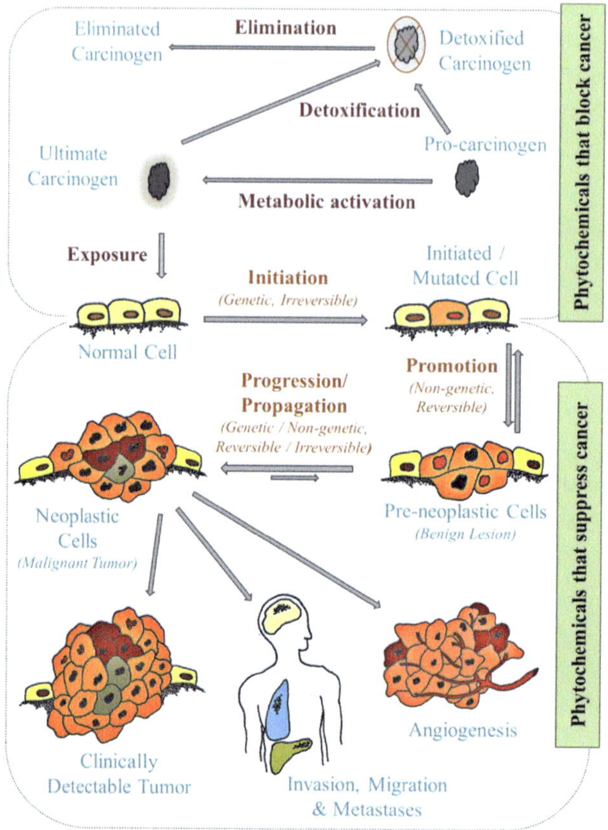

Figure 5.3. Anticancer mechanism of flaxseed lignans.

Source: De Silva SF, Alcorn J. Flaxseed Lignans as Important Dietary Polyphenols for Cancer Prevention and Treatment: Chemistry, Pharmacokinetics, and Molecular Targets. *Pharmaceuticals*. 2019; 12(2):68. https://doi.org/10.3390/ph12020068

The anti-cancer properties of milled flaxseed have been investigated through various methods including cell culture, animal experimentation, and human trials to elucidate the underlying mechanisms. According to certain theories, the presence of high concentrations of SDG and ALA in flaxseed is believed to be a contributing factor to its anticancer properties. According to a study, lignans, which encompass SDG, exhibit a reduction in overall mortality ranging from 40-53% and a decrease in breast cancer mortality ranging from 33-70%. According to a studythe administration of 50 mg SDG has demonstrated a reduction in the risk of cancer. Conversely, there is uncertainty regarding the direct attribution of anti-tumor activity to SDG.

The distribution of SDG is limited to the gastrointestinal tract and it does not exhibit systemic circulation. The process of breakdown into enterolactone and enterodiol occurs within the digestive tract. The aforementioned compounds exhibit characteristics of estrogenicity and antioxidation. Enterolactone and enterodiol can bind to estrogen receptors, which can consequently impact cell proliferation. It is hypothesized that the antitumor properties of SDG may be attributed to the SDG metabolites. The potential significance of anti-inflammatory properties of ALA has been suggested. Furthermore, the consumption of flaxseed oil, which is abundant in ALA, has been found to augment the antineoplastic properties of chemotherapy medications. The addition of flax oil to trastuzumab treatment for breast cancer patients expressing epidermal growth factor receptor 2 resulted in a noteworthy improvement in the antitumor effects of trastuzumab. Furthermore, the required doses of trastuzumab to eliminate tumors in athymic mice were significantly reduced, as evidenced by previous research. Comparable outcomes were noted upon the amalgamation of flaxseed and tamoxifen, an additional anti-neoplastic agent (Balducchi, 2023).

Flaxseed and its bioactive components have been found to impact various cellular oncotic pathways, including the regulation of estrogen metabolism, inhibition of cellular proliferation, angiogenesis, metastasis, and inflammation, as well as the induction of apoptosis within the tumor. The bioactive components present in flaxseed have been observed to or inhibit the phosphorylation of p-AKT, p-ERK, and p-JNK kinases, which leads to a deceleration of the MAPK pathways. Additionally, flaxseed has been found to inhibit CDK4, regulate multiple miRNAs, and reduce the expression of mRNAs for Bcl2, cell cycle proteins, ERalpha and beta, and epidermal and insulin-like growth receptors. These findings have been reported in various studies. A potential area for future research involves further exploration of the practical applications of dietary flaxseed, flax oil, and SDG, as well as their effectiveness in human clinical trials. There is a growing body of evidence indicating that flaxseed and its bioactive constituents may have beneficial effects on cancer prevention and treatment, similar to their positive impact on cardiovascular disease (Ranveer, 2015).

5.2.5 Relieves menopausal symptoms

Due to the estrogenic effects of its lignans, flaxseed has been proposed as a viable alternative to hormone replacement treatment or as an additional strategy for hormone management (Chang et al., 2019). According to research, flaxseed lignans can benefit menopausal women by reducing their risk of heart disease. By preserving hormone homeostasis, flaxseed may reduce the risk of developing osteoporosis. Additionally, according to some research, it may be able to lessen mild to severe menopausal symptoms including mood swings, hot flashes, and vaginal dryness

A flaxseed diet may potentially have a preventative impact in addition to protecting against menopausal symptoms. Numerous studies have looked at the impact of flaxseed or its bioactive components on postmenopausal women's quality of life, frequency, and severity of hot flashes. It has been suggested that the estrogenic action of several flaxseed metabolites may have a positive impact on a variety of postmenopausal symptoms. Those who ingested flaxseed reported fewer menopausal symptoms and an improvement in their general quality of life, according to research done on a sample of 140 postmenopausal women. A big dose of flaxseed (40 g per day) was found to effectively reduce menopausal symptoms in research with a sizable sample size of 199 women and a longer duration of 1 year. However, it was shown that this impact was equivalent to that of the control group, who received a wheat germ placebo. Another study found conflicting evidence of flaxseed's beneficial effects on menopausal symptoms, with a daily consumption limit of 90 g. Flaxseed did not significantly affect menopausal hot flashes or quality of life, according to the results of randomized controlled studies and clinical trial reviews. Randomized, placebo-controlled trials must be conducted to produce conclusive data on the influence on symptoms related to menopause. Based on the results of controlled trials, it can be concluded that flaxseed does not affect menopausal women in terms of the quality of life or the frequency and intensity of hot flashes (Balducchi, 2023).

Flaxseed consumption has been advised to be avoided during pregnancy and breastfeeding. In rats that were pregnant, the effects of SDG and flaxseed oil on body fat mass were shown to be completely different. Administration of the SDGs was discovered to have a considerable influence on the growth of the fat mass, but flaxseed oil ingestion was found to have a moderating impact. Additionally, research indicated that breastfeeding rodents given flax oil showed reduced levels of triglycerides and cholesterol in both their blood and milk. According to research investigations, both the male and female offspring of moms who took flaxseed oil showed lower levels of triglycerides and body fat mass. Given the metabolic differences between the systems of the mother and the fetus, it would be prudent to use caution while taking flaxseed or its components during pregnancy. As no other research has observed similar alterations, randomized, placebo-controlled trials including pregnant women are required to demonstrate solid proof.

Women who have had menopause are more likely to develop osteoporosis. postmenopausal women who consumed flaxseed had somewhat lower levels of tartrate-resistant acid phosphatase, a marker of bone resorption. However, anabolic bone medications had little of an effect. However, only a small number of studies have looked at the effects of a flaxseed-rich diet on bone mineral density, bone mineral content, and bone turnover in menopausal women.

The simultaneous administration of flaxseed and estrogen treatment had an extra positive effect in an animal model. It has been shown that ALA-rich flaxseed oil consumption is advantageous for bone health, particularly in situations of pathological diseases like obesity and renal illness. ALA, not flaxseed's estrogenic lignan concentration, may have contributed to the improvement in osteoporosis-related bone problems that have been seen (Ranveer, 2015).

5.2.6 Reduce the risk of osteoporosis

Flaxseed helps maintain bone density and strength while preventing bone loss. Consumption of flaxseed oil increases bone mineral density and decreases osteoporosis risk factors. Women with diabetes can reduce their risk of osteoporosis by consuming flaxseed oil in their diets.

5.2.7 The gluten-free grain

Flaxseed presents a viable gluten-free substitute in recipes that require wheat, rye, or barley flour. Grains, especially those that contain gluten, can elicit considerable digestive challenges and adverse reactions for a substantial number of people. The incorporation of flaxseed granules into various food formulations has the potential to impart gluten-like properties. Flaxseed granules possess the ability to effectively absorb substantial amounts of liquid and serve as a binding agent in various baking and culinary preparations. Individuals diagnosed with celiac disease or exhibiting gluten sensitivity are advised to opt for flaxseed due to its gluten-free nature.

5.2.8 High in antioxidants

Flaxseed is rich in antioxidants, especially lignans, which are polyphenols associated with the insoluble flaxseed fiber. Antioxidants in foods, such as lignan, mitigate the damage caused by free radicals in a variety of diseases and conditions (Balducchi, 2023).

5.2.9 Healthy skin and hair

As a medical substance, flaxseed oil reduces skin irritation reactions such as stinging, dermatitis, and edema. It has also been reported that flaxseed oil and gum formulations can aid in preventing hair loss. High levels of B vitamins and omega-3 fatty acids (ALA) in flaxseed aid to reduce skin dehydration and flakiness. The anti-inflammatory properties of flaxseed oil assist in the treatment of a variety of dry skin conditions, including acne, psoriasis, and atopic dermatitis.

According to reports, flaxseed oil is more effective against parched skin than whole-toasted flaxseed against inflammation. Some studies indicate that the high ALA content of flaxseed oil can prevent the epidermis from drying out. It is recommended to use an emollient with flaxseed oil and essential oils or to consume 10 mL of flaxseed oil daily. Flaxseed mucilage is a crucial component of treatments that prevent hair loss. This component consists of a similar substance to mucin and pectin. Due to the similarity between flaxseed pectin and mucin, a combination is formed that adheres to the cranium and eliminates dead skin cells without creating irritation. Mucin prevents the breakdown of proteins, thereby protecting the hair cuticle from damage and devastation. Mucin and flaxseed gum have comparable effects on the hair cuticle. Additionally, the flaxseed protein conlinin can help to soften hair, repair damaged hair, and moisturize a parched scalp and hair. Flaxseed is commonly used in hair gel formulations as an emollient to preserve hair resilience (Ranveer, 2015).

5.2.10 Weight loss

Multiple randomized controlled trials have demonstrated that flaxseed can aid in weight loss. Supplementation with flaxseed was most beneficial for individuals with metabolic syndrome, rotund adults with pre-diabetes, and polycystic ovary syndrome. Extreme inflammation increases visceral fat, whereas pro-inflammatory cytokines decrease adiponectin, a hormone produced by fat cells that prevent weight gain. Therefore, the mechanism of flaxseed consumption may be related to its anti-inflammatory properties. Men who followed a weight loss diet containing flaxseed lost weight and had lower levels of inflammatory biomarkers, such as C-reactive protein (CRP) and tumor necrosis factor-alpha (TNF-), in comparison to a control group administered a comparable diet without flaxseed. Additionally, flaxseeds are rich in fiber, which promotes satiety after consumption. By lowering the desire to eat, prolonged satiety contributes to the fight against obesity. With a nutritious diet and 40 grams of pulverized flaxseed per day, weight loss is facilitated.

5.2.11 Dietary flaxseed and cardiovascular disease

The impact of flaxseed consumption on risk factors associated with cardiovascular disease is a commonly discussed topic in the realm of health benefits. Cardiovascular disease accounts for 33% of all deaths in the United States. According to statistical data, cardiovascular disease causes the demise of more than 2,200 individuals in the United States daily, which equates to one fatality every 40 seconds. The discovery of novel therapeutic targets has expedited the progress of pharmacological interventions that target specific pathways associated with cardiovascular disorders. Despite this advancement, these therapies are still constrained by limitations in patient compliance and adverse effects. Plant-based diets possess functional nutrients that could potentially be utilized for the prevention or treatment of cardiovascular disease. Studies, encompassing both preclinical and clinical trials, are being carried out to gain further insight into the impact of functional foods on cardiovascular well-being (Balducchi, 2023).

Several clinical studies have provided evidence that n-3 polyunsaturated fatty acids (PUFAs) possess cardioprotective properties. Several studies have demonstrated that n-3 PUFAs offer protection against cardiovascular diseases and associated risk factors, including hypertension, platelet aggregation, dyslipidemia, atherosclerosis, and arrhythmias. The human diet predominantly relies on fish as the primary source of PUFAs. Numerous research endeavors have established a correlation between the intake of fish oil and a noteworthy decrease in risk factors associated with cardiovascular disease. The utilization of fish oil is limited due to apprehensions regarding its sensory attributes, including flavor and odor, as well as potential hypersensitivity reactions and the existence of contaminants such as methylmercury, dioxins, and polychlorinated biphenyls, despite its established cardiovascular advantages (Ranveer, 2015).

Studies have demonstrated that the intake of flaxseed can decelerate the advancement of atherosclerosis in animal models of cardiovascular disease that were induced by the consumption of high levels of cholesterol or trans-fat. The observed phenomenon is believed to be attributed to the anti-inflammatory properties of ALA present in flaxseed, as reported in previous studies. The effect of flaxseed on circulatory cholesterol levels varies depending on the animal species, as evidenced by studies conducted on rodents and rats indicating a potential decrease, while studies on rabbits suggest no significant impact. These findings have been documented in sources. According to a study, flaxseed has the potential to assist in reducing levels of trans fats in the bloodstream. Research has demonstrated that flaxseed supplements possess the ability to reverse atherosclerosis in stable and established atherosclerotic plaques.

Arrhythmias may arise in isolated hearts when they are subjected to ischemia-reperfusion. According to a study, rabbits that were administered a diet enriched with flaxseed before the ischemia-reperfusion insult was able to prevent ventricular fibrillation. According to a study, rats that were provided with a diet enriched with flaxseed exhibited reduced infarct size, fewer incidences of arrhythmias, and lesser left ventricular dilatation both before and during the onset of myocardial infarction. The ALA content present in flaxseed is presumed to have a significant impact on the aforementioned process. Research has shown that ALA can modify ionic cardiac currents which can lead to arrhythmias. The transformation of SDG present in the circulation into enterolignans is believed to possess a potential cardioprotective impact, as suggested by previous research studies (Balducchi, 2023).

Research has shown that the consumption of flaxseed provides significant benefits for the prevention of cardiovascular disease in individuals who are at risk. A noteworthy observation is the reduction in both systolic and diastolic blood pressure among individuals diagnosed with peripheral artery disease (PAD). The findings of this research indicate that the consumption of flaxseed resulted in a significant reduction in both brachial and central blood pressure, as reported in previous studies. The FlaxPAD Trial was a randomized, double-blind, and placebo-controlled study. The study involved patients with peripheral artery disease who were administered 30 g of milled flaxseed daily for six months. The results indicated a significant decrease in both systolic and diastolic blood pressure. During the six months, it was observed that 21 individuals in the placebo group exhibited an elevation in systolic blood pressure (SBP), whereas 15 individuals demonstrated a reduction in SBP. During the six months, a reduction in systolic blood pressure was observed in 27 patients who were administered flax, whereas an increase in systolic blood pressure was observed in 12 patients who were not administered flax (control group) (Ranveer, 2015).

Throughout this temporal interval, the systolic blood pressures (SBPs) of two persons remained constant. The flax group exhibited a noteworthy and statistically significant mean reduction of 10 mm Hg. The study reported a hypotensive effect of flaxseed in the diet, which was observed within a month and persisted for one year. It was anticipated that the reduction would result in a 50% decrease in the occurrence of myocardial infarctions and strokes. The aforementioned intervention resulted in a decrease in the synthesis of oxylipins that exhibit diminished vasodilatory properties and heightened pro-inflammatory effects. This outcome was accompanied by a notable alteration in lipid metabolism, as reported in the reference. As per a particular hypothesis, the occurrence of this phenomenon can be attributed to the ALA composition present in flaxseed. Studies conducted on animals have demonstrated that the application of flaxseed protein hydrolysates or flax oil, whether in acute or chronic doses, resulted in a reduction of blood pressure in spontaneously hypertensive rats. Furthermore, it was observed that a concentrate of flaxseed lignan caused a reduction in blood pressure among rodents that were provided with high-fat, high-sugar diets. Similar results were also observed in animals that were administered deoxycorticosterone acetate (DOCA-salt) to induce renal hypertension.

5.2.12 Dietary flaxseed and the brain

It is well known that omega-3 fatty acids from marine sources play an essential role in brain development. Previously, the primary emphasis of this research was on the omega-3 fatty acids docosahexaenoic acid (DHA) and eicosapentaenoic acid (EPA). EPA appears to influence behavior and temperament, but research indicates that DHA is crucial for prenatal and postnatal brain development. A DHA deficiency has been linked to impaired learning and spatial performance. The omega-3 fatty acid ALA, which is abundant in flaxseed, may also benefit the brain, according to these findings. During pregnancy, rats fed flaxseed gave birth to offspring with enlarged brains that were significantly enriched in ALA and DHA. Puppies fed milled flaxseed or flaxseed oil shortly after birth had greater brain mass, suggesting the importance of milled flaxseed in particular for promoting early postnatal brain development. Nonetheless, research indicates that the diet may inhibit body growth due to an imbalance in omega-3 and omega-6 fatty acid levels, whereas flaxseed consumption may enhance brain development and spatial memory (Ranveer, 2015).

In the presence of a neurological disorder, flaxseed consumption may enhance certain aspects of brain function. Consumption of flaxseed by the mother offered neuroprotection during experimental neonate hypoxic-ischemic encephalopathy by reducing brain mass loss and enhancing motor hyperactivity and spatial memory. It also prevented the development of depressive symptoms in adolescents. Consumption of flaxseed reduced all symptoms of chronic stress in rodents exposed to moderate stress over time. It has been demonstrated that flaxseed lignans, such as SDG, have antidepressant-like effects in rodents under protracted stress and neuroprotective properties in cortical neurons against NMDA-induced neurotoxicity (Balducchi, 2023).

Chapter

6 BIOACTIVITY OF ALKYLRESORCINOLS

CONTENTS

6.1 INTRODUCTION

The bioactivity of natural compounds has garnered significant attention in recent years due to their potential therapeutic applications. Alkylresorcinols, a class of phenolic compounds, have been of particular interest due to their diverse biological activities.

Alkylresorcinols (ARs) are natural compounds found in various cereal products such as wheat, rye, and triticale. They possess a range of physiological and biological activities, including bactericidal, antioxidant, enzyme activity modulation, and tumor inhibition. The bioactivity of ARs is closely related to their absorption and metabolism in the body, with the small intestine playing a role in their partial absorption or metabolism. ARs have been detected in plasma at concentrations ranging from 1.47 to 124 µg/L, indicating their presence and potential effects in the bloodstream (Zhao, 2020).

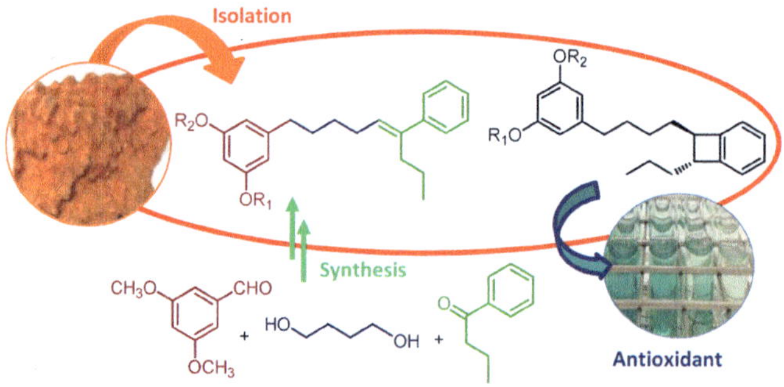

Figure 6.1. Alkylresorcinol.

Source: Ortega MJ, Pantoja JJ, De los Reyes C, Zubía E. 5-Alkylresorcinol Derivatives from the Bryozoan *Schizomavella mamillata*: Isolation, Synthesis, and Antioxidant Activity. *Marine Drugs*. 2017; 15(11):344. https://doi.org/10.3390/md15110344

The alkyl chain length of ARs is a crucial factor influencing their biological activity, primarily due to its impact on fat solubility. Longer alkyl chains tend to exhibit better fat solubility. Different ARs have been identified in cereal products, including tricosylresorcinol (C23:0), pentacosylresorcinol (C25:0), nonadecylresorcinol (C19:0), heptadecylresorcinol (C17:0), and heneicosylresorcinol (C21:0). These compounds vary in their alkyl chain lengths, molecular weights, and structures

Structurally, Alkylresorcinols (ARs) are natural compounds that have been found in various plant families, including *Anacardiaceae, Ginkgoaceae, Proteaceae, Myrsinaceae, Primulaceae, Myristicaceae, Iridaceae, Compositae, Leguminosae,* and *Gramineae.* The amount of ARs present in plants varies significantly, ranging from trace amounts to relatively high concentrations. For example, *Hordeum vulgare* (barley) grains may contain ARs as low as 0.005% (m/m), while *Secale cereale* (rye) grains can contain up to 0.3% (m/m) of ARs. However, the highest concentration of ARs has been reported in the oil extract of cashew nut shells, reaching up to 20% (m/m) (Ross, 2012).

One distinctive feature of ARs is their amphiphilic nature, characterized by the presence of both a water-soluble polar "head" (dihydroxybenzene group) and a water-insoluble non-polar alkyl "tail." This unique molecular structure has significant implications for the analysis, absorption, metabolism, and potential bioactivity of ARs. The hydrophobicity of ARs increases with the length of the alkyl chain, as indicated by octanol/water partition coefficients (log Po/w) ranging from 7.0 to 13.4 (Zhao, 2020).

While ARs can be isolated from various sources, they offer the greatest dietary advantage when derived from edible food sources. Whole-grain rye or wheat kernels are particularly important dietary sources of ARs. Research has extensively explored the potential positive effects of ARs on human health, highlighting their significance in the human diet.

However, the interest in ARs initially arose due to concerning findings in the livestock industry, suggesting that ARs present in cattle feed could inhibit growth and decrease appetite in animals. Consequently, there was a need to develop a rapid method for detecting the presence of ARs. Over time, research has shifted to exploring the potential bioactivity of ARs in humans, indicating promising outcomes.

ARs are known to be present during the development of cereal kernels but decrease in quantity as the grains mature. This decrease might be attributed to a dilution effect caused by the increase in starch and protein proportions during grain maturation.

Notably, ARs exhibit protective properties for plants, particularly against mold and external harmful organisms like bacteria. Their localization in the outer layers, such as the bran of cereal grains, enhances their protective effects. The intermediate layer between the testa and pericarp has been identified as the primary site of AR accumulation, containing approximately 99% of the total ARs in a cereal kernel.

This chapter aims to evaluate the bioactivity of ARs in different cereal products, focusing on their antibacterial activity, antioxidant properties, anticancer potential, and inhibition of enzymes (Ross, 2012).

6.2 BIOSYNTHESIS OF ALKYLRESORCINOLS

Alkylresorcinols are a type of phenolic lipids that are synthesized through the polyketide pathway, specifically involving an enzyme called 2'-oxoalkylresorcylic acid synthase (ORAS). The polyketide pathway is one of the main pathways for the biosynthesis of plant phenolic compounds, alongside the shikimate pathway.

The biosynthesis of alkylresorcinols starts with the formation of polyketomethylenic chains. These chains are derived from acetic acid and activated forms of acetyl-S-CoA and malonyl-S-CoA. This process is similar to the biosynthesis of fatty acids, as both pathways involve the addition of C2 units to form linear chains (Zhao, 2020).

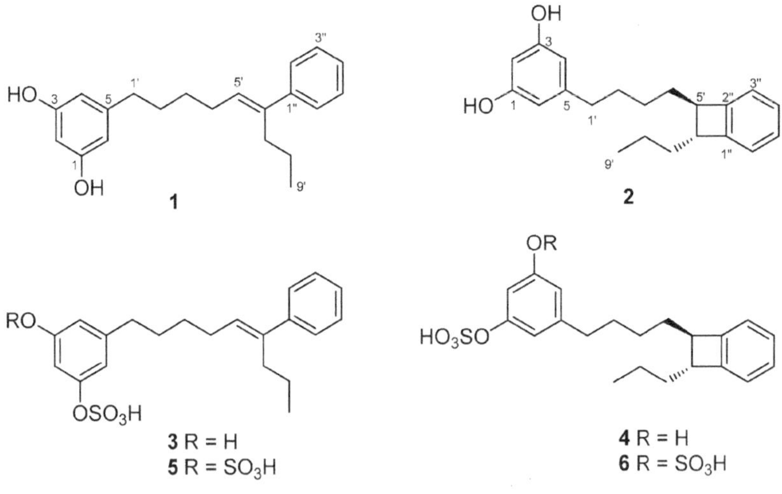

Figure 6.2:5. Alkylresorcinol Derivatives from the Bryozoan *Schizomavella mamillata.*

Source: Ortega MJ, Pantoja JJ, De los Reyes C, Zubía E. 5-Alkylresorcinol Derivatives from the Bryozoan *Schizomavella mamillata*: Isolation, Synthesis, and Antioxidant Activity. *Marine Drugs*. 2017; 15(11):344. https://doi.org/10.3390/md15110344

In the case of alkylresorcinols, the alkyl side chain, which is an odd-numbered side chain, is a direct result of the fatty acid precursor. The biosynthesis pathway proceeds through several steps. A fatty acyl-CoA molecule reacts with malonyl coenzyme A three times through an aldol condensation-type reaction. This series of reactions results in the formation of a tetraketide intermediate.

Next, the tetraketide intermediate undergoes an aldol condensation-type ring closure and hydrolysis, releasing the coenzyme A. This step leads to the formation of alkylresorcylic acid. Finally, a non-enzymatic decarboxylation occurs, resulting in the production of 5-n-alkylresorcinols, which are the final products.

The enzyme 2'-oxoalkylresorcylic acid synthase (ORAS) plays a critical role in the biosynthesis of alkylresorcinols. ORAS is responsible for synthesizing various compounds, including triketide and tetraketide pyrones, as well as pentaketide resorcylic acids and resorcinols, from the fatty acyl-CoA starter unit (Ross, 2012).

Overall, the biosynthesis of alkylresorcinols involves the polyketide pathway, with ORAS playing a crucial role in the synthesis of the necessary intermediates and final products. This pathway utilizes acetic acid, Acetyl-S-CoA, and malonyl-S-CoA to generate polyketomethylenic chains that undergo ring closure, hydrolysis, and decarboxylation to produce alkylresorcinols.

The biosynthesis of alkylresorcinols showcases the intricate and diverse nature of plant secondary metabolite synthesis. The polyketide pathway, in conjunction with specific enzymes like ORAS, allows plants to produce a wide range of phenolic compounds with varying structures and biological activities. Understanding the biosynthetic pathways of these compounds provides insights into their ecological functions and potential applications in various industries, including pharmaceuticals, cosmetics, and food science (Zhao, 2020).

The biosynthesis of alkylresorcinols is a complex process that occurs through the polyketide pathway. This pathway, facilitated by enzymes like 2'-oxoalkylresorcylic acid synthase (ORAS), utilizes acetic acid, acetyl-S-CoA, and malonyl-S-CoA to generate polyketomethylenic chains that undergo various enzymatic reactions, leading to the formation of alkylresorcinols.

However, various methods of solvent extraction are commonly employed to extract alkylresorcinols (ARs) from cereals using organic solvents like acetone, ethyl acetate, and methanol. However, when it comes to extracting ARs from bread, acetone and ethyl acetate are not suitable. Instead, a hot

1-propanol: water (3:1, v/v) solution is utilized to ensure the complete recovery of ARs from baked products. Initially, it was believed that baking destroyed or reduced the ARs content in bread, as only small quantities were detected. However, subsequent experiments demonstrated that ARs remain stable during baking and food processing.

ARs are amphiphilic compounds that become less soluble in water as the alkyl chain length increases. Consequently, organic solvents such as acetone and ethyl acetate have traditionally been used to extract ARs from cereal grains at room temperature for 16-24 hours, with a sample-to-organic solvent ratio ranging from 1:40 to 1:50 (w/v). Nonetheless, the organic solvent extraction methods possess certain drawbacks, including lengthy extraction times, generation of toxic waste, and lack of selectivity. Moreover, concerns regarding food security, safety, and quality arise when employing organic solvents.

An alternative to traditional solvent extraction methods is supercritical carbon dioxide (SC-CO_2) extraction, which can be utilized for the biosynthesis and extraction of alkylresorcinols (Ross, 2012).

6.3 ANTIMICROBIAL ACTIVITY

One of the most well-studied bioactivities of alkylresorcinols is their antimicrobial properties. Several studies have demonstrated their inhibitory effects against a wide range of bacteria, fungi, and viruses. The mechanism of action involves disruption of microbial cell membranes, leading to cell lysis and subsequent death. Alkylresorcinols have shown promise as natural antimicrobial agents in various applications, such as food preservation and pharmaceutical formulations.

6.4 ANTI-INFLAMMATORY AND ANTIOXIDANT EFFECTS

Alkylresorcinols exhibit significant anti-inflammatory and antioxidant activities, making them potential candidates for the treatment of inflammatory diseases and oxidative stress-related conditions. These compounds have been shown to suppress the production of pro-inflammatory mediators and inhibit the activity of enzymes involved in inflammation. Additionally, alkylresorcinols act as free radical scavengers, protecting cells from oxidative damage (Zhao, 2020).

Oxidative damage caused by reactive oxygen species (ROS) and reactive nitrogen species (RNS) is a well-known phenomenon that contributes to various inflammatory and degenerative diseases. Free radicals, classified as ROS and RNS, are byproducts of aerobic metabolism and can lead to cell damage when there is an imbalance between oxidants and antioxidants. The detrimental effects of free radicals have been linked to conditions such as diabetes, atherosclerosis, and other diseases.

Dietary antioxidants play a crucial role in counteracting the harmful effects of free radicals. Phenolic antioxidants are a class of compounds that are particularly effective in inhibiting the oxidation process. They act by various mechanisms, including acting as physical barriers, chemical traps, catalytic systems, and metal ion binders, which prevent or neutralize the generation of ROS. For example, carotenoids, anthocyanins, and other phenolic compounds serve as chemical traps, absorbing energy and electrons to quench ROS. Antioxidant enzymes like superoxide dismutase, catalase, and glutathione peroxidase function as catalytic systems to neutralize ROS. Additionally, antioxidants such as ascorbic acid, tocopherols, and flavonoids act as chain-breaking antioxidants, preventing the propagation of lipid peroxidation (Ross, 2012).

The oxidation process can occur through two pathways: H-atom transfer (HAT) and single electron transfer (SET). Antioxidants can play a preventive role in both pathways. In the HAT pathway, an antioxidant interrupts the chain reaction by reacting with free radicals and hydroxyperoxides as shown below:

- $RH \rightarrow R^{\cdot}$ (initiation)
- $R^{\cdot} + O2 \rightarrow ROO^{\cdot}$ (addition of O2)
- $ROO^{\cdot} + RH \rightarrow ROOH + R^{\cdot}$ (H-atom transfer)
- $ROO^{\cdot} + AH \rightarrow ROOH + AO^{\cdot}$

The effectiveness of an antioxidant depends on factors such as stability as a free radical, slow reaction with the substrate, and rapid reaction with hydroxyperoxides. One of the most effective lipid-soluble chain-breaking antioxidants in human blood plasma is α-tocopherol, the active component of Vitamin E.

The efficacy of phenolic antioxidants is determined by their bond dissociation enthalpy (BDE), with weaker OH bonds reacting faster with free radicals. Studies have shown that alkylresorcinols (ARs), derived from cereal bran isolates, exhibit antioxidant activity. Kozubek and Nienartowicz found that ARs with shorter alkyl chain lengths had higher antioxidant activity, with AR-C15:0 showing the highest effect followed by C19:0 and C23:0. Another study by Winota and Lorenz demonstrated that AR homolog-C15:0 effectively slowed the rate of oxidative rancidity in oil and lipid oxidation in cereal products.

The antioxidant activity of ARs in food and biological systems has been established, but there is still a need for further research to fully understand their mechanisms of action. Detailed studies are required to investigate how ARs act in membranes, different emulsions, or food systems, their potential as preventive antioxidants, and their potential synergistic effects with other compounds (Zhao, 2020).

6.5 PREBIOTIC ACTIVITY

Dietary fiber (DF) encompasses cell wall polysaccharides, lignin, and other substances that resist digestion by human enzymes, as defined by the American Association of Cereal Chemists (AACC). Among these resistant compounds are dietary antioxidants, including phenolic acids such as ferulic acid, sinapic acid, p-coumaric acid, and caffeic acid, with ferulic acid being the most abundant in wheat. These phenolic compounds are predominantly bound to cell wall polysaccharides, forming ester bonds. For example, ferulic acid is attached to arabinoxylans through acylation of the primary hydroxyl group at the C5 position of α-arabinofuronasyl residues.

The hydrolysis of these ester bonds can occur through wheat-related endoxylanases or specific intestinal bacteria that possess arabinoxylan-degrading enzymes. This hydrolysis process leads to the formation of arabinoxylan oligosaccharides (AXOS), which are considered potential prebiotics. AXOS represents a new class of candidate prebiotics; however, limited information is available regarding their prebiotic potency *in vivo* (Ross, 2012).

The bioactivity of alkylresorcinols, a group of compounds found in wheat and rye, has been extensively studied, with a focus on their free radical scavenging and antioxidant properties. Alkylresorcinols have been shown to possess significant antioxidant activity, as they can neutralize free radicals and prevent oxidative damage in various biological systems. In particular, studies have demonstrated the ability of alkylresorcinols to scavenge reactive oxygen species (ROS), such as superoxide anions and hydroxyl radicals, which are known to contribute to cellular oxidative stress.

Research has also indicated that alkylresorcinols can protect against lipid peroxidation, a process that leads to the production of harmful lipid peroxides. For instance, in a study by Andersson et al., it was observed that alkylresorcinols extracted from whole-grain wheat bran effectively inhibited lipid peroxidation in a dose-dependent manner in a model system using rat liver microsomes. The researchers suggested that the antioxidant activity of alkylresorcinols may be attributed to their ability to donate hydrogen atoms to lipid peroxyl radicals, thereby breaking the chain reaction of lipid peroxidation.

Furthermore, alkylresorcinols have been shown to exhibit anti-inflammatory effects by inhibiting the production of pro-inflammatory cytokines and enzymes, such as interleukin-6 (IL-6) and cyclooxygenase-2 (COX-2). In a study conducted by Li et al., it was found that alkylresorcinols derived from rye bran suppressed the expression of IL-6 and COX-2 in lipopolysaccharide-stimulated macrophages, suggesting their potential role in modulating inflammatory responses (Zhao, 2020).

In addition to their direct antioxidant and anti-inflammatory activities, alkylresorcinols have been implicated in modulating gut microbiota composition and promoting the growth of beneficial bacteria. Studies have shown that alkylresorcinols can act as prebiotics by selectively stimulating the growth of Lactobacilli and Bifidobacteria, which are known to have positive effects on gut health. For instance, it was shown in a study by Rinnan et al. that alkylresorcinols from entire whole-grain wheat encouraged the growth of Bifidobacterium spp. and Lactobacillus spp. in a model of simulated colonic fermentation. Moreover, the metabolites produced by the gut microbiota during the metabolism of alkylresorcinols may also contribute to their bioactivity. These metabolites, such as phenolic acids, short-chain fatty acids, and other microbial-derived compounds, have been reported to possess various health benefits, including antioxidant and anti-inflammatory effects. However, it is important to note that most of the current evidence regarding the bioactivity of alkylresorcinols is based on *in vitro* and animal studies and further research is needed to confirm these effects in humans (Ross, 2012).

Alkylresorcinols derived from wheat and rye exhibit promising bioactivity, particularly in terms of their free radical scavenging and antioxidant properties. These compounds have demonstrated the ability to neutralize free radicals, protect against lipid peroxidation, modulate inflammatory responses, and promote the growth of beneficial gut bacteria.

The potential health benefits of alkylresorcinols warrant further investigation to establish their significance in human health and disease prevention.

6.6 ANTI-CANCER PROPERTIES

Emerging evidence suggests that alkylresorcinols possess anti-cancer properties. Studies have revealed their ability to inhibit tumor growth and induce apoptosis (programmed cell death) in various cancer cell lines. The underlying mechanisms involve modulation of cell signaling pathways, interference with DNA replication, and disruption of mitochondrial function. Alkylresorcinols hold promise as adjunctive therapeutic agents in cancer treatment, although further research is needed to fully elucidate their potential (Zhao, 2020).

One of the primary mechanisms through which alkylresorcinols exhibit anti-cancer properties is their ability to induce apoptosis, or programmed cell death, in cancer cells. Apoptosis plays a crucial role in maintaining the balance between cell proliferation and cell death, and dysregulation of this process contributes to the development and progression of cancer. Alkylresorcinols have been shown to activate specific pathways that promote apoptosis in cancer cells while sparing healthy cells. For example, a study conducted by Xie et al. (2018) demonstrated that a specific alkylresorcinol compound extracted from wheat bran effectively induced apoptosis in breast cancer cells by activating caspase-dependent pathways. This finding highlights the potential of alkylresorcinols as targeted anti-cancer agents.

Moreover, alkylresorcinols have shown promise in inhibiting angiogenesis, the process by which new blood vessels are formed to supply tumors with nutrients and oxygen. By targeting the factors and signaling pathways involved in angiogenesis, alkylresorcinols can potentially halt tumor growth and metastasis. For instance, research conducted by Masoodi et al. (2020) revealed that a specific alkylresorcinol compound derived from sesame oil exhibited anti-angiogenic effects by inhibiting the vascular endothelial growth factor (VEGF) signaling pathway in colorectal cancer. This inhibition disrupted the formation of new blood vessels, ultimately impeding tumor progression (Ross, 2012).

Furthermore, alkylresorcinols have been shown to possess potent antioxidant properties. Oxidative stress, characterized by an imbalance between reactive oxygen species (ROS) production and antioxidant defense mechanisms, plays a critical role in cancer development and progression. Alkylresorcinols act as free radical scavengers, neutralizing ROS and reducing oxidative stress levels. For instance, a study conducted by Insaf et al. (2022) demonstrated that an alkylresorcinol compound derived from the marine brown alga Ecklonia cava exhibited strong antioxidant activity by suppressing ROS generation and lipid peroxidation. This antioxidant effect may help protect cells from DNA damage and mutations that can lead to cancer initiation.

In addition to their direct anti-cancer effects, alkylresorcinols have been found to enhance the efficacy of conventional anti-cancer therapies. Combination treatments involving alkylresorcinols and chemotherapy drugs have shown synergistic effects, leading to improved treatment outcomes. For example, a study by Nova et al. (2022) investigated the combination of an alkylresorcinol compound derived from bee pollen with the chemotherapy drug cisplatin in lung cancer cells. The study found that the combined treatment resulted in enhanced cancer cell death compared to cisplatin alone. This suggests that alkylresorcinols have the potential to augment the effectiveness of existing anti-cancer therapies, reducing the required drug dosage and minimizing adverse side effects.

Perhaps, alkylresorcinols demonstrate their potential as effective anti-cancer agents through their ability to induce apoptosis, inhibit angiogenesis, exhibit antioxidant properties, and enhance the efficacy of conventional therapies. These may include neurodegenerative illness therapeutics (Zhao, 2020).

6.7 NEUROPROTECTIVE EFFECTS

Neurodegenerative diseases, such as Alzheimer's and Parkinson's disease, pose significant challenges in healthcare. The bioactivity of alkylresorcinols, specifically their neuroprotective properties, has been the subject of extensive research and evaluation in recent years. Alkylresorcinols are a group of naturally occurring phenolic compounds commonly found in various plant species, including cereals such as wheat, rye, and barley. These compounds have gained attention due to their potential therapeutic applications in the field of neuroprotection.

One of the primary mechanisms through which alkylresorcinols exert their neuroprotective effects is by acting as antioxidants. Oxidative stress, caused by an imbalance between the production of ROS and the body's antioxidant defense mechanisms, plays a crucial role in the development and progression of various neurodegenerative disorders, such as Alzheimer's disease and Parkinson's disease. Alkylresorcinols possess the ability to scavenge ROS and neutralize their harmful effects, thereby mitigating oxidative stress-induced damage to neurons (Ross, 2012).

Furthermore, alkylresorcinols have been shown to possess anti-inflammatory properties, which can contribute to their neuroprotective effects. Chronic inflammation has been implicated in the pathogenesis of neurodegenerative disorders, and controlling inflammatory processes in the brain is crucial for maintaining neuronal health. Alkylresorcinols can modulate inflammatory responses by inhibiting the production of pro-inflammatory cytokines and reducing the activation of inflammatory signaling pathways.

A notable example highlighting the anti-inflammatory properties of alkylresorcinols is a study by Vadim et al. (2020) that explored the effects of these compounds on neuroinflammation in an *in vitro* model of Alzheimer's disease. The researchers demonstrated that alkylresorcinols extracted from barley bran effectively suppressed the release of inflammatory mediators, such as interleukin-1β and tumor necrosis factor-alpha, from microglial cells. This suggests that alkylresorcinols may attenuate neuroinflammation, thereby protecting neurons from inflammatory damage associated with Alzheimer's disease.

In addition to their antioxidant and anti-inflammatory activities, alkylresorcinols have been found to possess other neuroprotective properties, such as promoting neuronal survival and enhancing neuronal plasticity. These compounds have been shown to modulate various signaling pathways involved in neuronal growth and survival, including the PI3K/Akt and MAPK/ERK pathways. By activating these pathways, alkylresorcinols can enhance the resilience of neurons and promote their adaptation to stressors.

For instance, a study by Wagner et al. (2019) investigated the neuroprotective effects of alkylresorcinols derived from rye bran in a cellular model of cerebral ischemia, a condition characterized by a lack of blood flow to the brain. The researchers demonstrated that alkylresorcinols promoted neuronal survival by activating the PI3K/Akt pathway, which plays a crucial role in cell survival and growth. Moreover, alkylresorcinols enhance the expression of neurotrophic factors, such as brain-derived neurotrophic factor (BDNF), which are essential for neuronal plasticity and regeneration.

Overall, the evaluation of the bioactivity of alkylresorcinols, with a particular focus on their neuroprotective properties, has revealed promising therapeutic potential. Through their antioxidant, anti-inflammatory, and pro-survival effects, alkylresorcinols can offer protection to neurons, mitigate oxidative stress and inflammation, and promote neuronal adaptation and plasticity (Zhao, 2020).

6.8 OTHER BIOACTIVITIES

In addition to the aforementioned bioactivities, alkylresorcinols have demonstrated other beneficial effects. These include anti-diabetic properties by regulating glucose metabolism, anti-obesity effects by inhibiting adipogenesis, and potential cardioprotective effects by modulating lipid metabolism and reducing cholesterol levels. These diverse bioactivities highlight the multifaceted therapeutic potential of alkylresorcinols (Ross, 2012).

6.8.1 Anti-diabetic activity

Alkylresorcinols have shown promise in regulating glucose metabolism, making them potentially useful in managing diabetes. Glucose metabolism refers to the process by which the body breaks down carbohydrates into glucose and utilizes it for energy. Diabetes is a chronic metabolic disorder characterized by high blood glucose levels, which can lead to various complications if left uncontrolled. Therefore, finding natural compounds that can effectively regulate glucose metabolism is of great importance in managing diabetes and its associated complications.

Several studies have investigated the anti-diabetic properties of alkylresorcinols and have shown promising results. These compounds have been found to exert their effects through multiple mechanisms, targeting various pathways involved in glucose regulation. One of the key mechanisms by which alkylresorcinols demonstrate their anti-diabetic activity is through enhancing insulin sensitivity.

Insulin is a hormone produced by the pancreas that plays a vital role in regulating glucose metabolism. In individuals with type 2 diabetes, cells become resistant to the actions of insulin, leading to impaired glucose uptake and utilization. Alkylresorcinols have been shown to enhance insulin sensitivity by promoting the translocation of glucose transporters, such as GLUT4, to the cell surface. This facilitates the uptake of glucose from the bloodstream into the cells, thereby reducing blood glucose levels.

Moreover, alkylresorcinols have been reported to inhibit key enzymes involved in glucose metabolism, such as α-amylase and α-glucosidase.

These enzymes are responsible for breaking down complex carbohydrates into simpler forms, such as glucose, which can be absorbed by the body. By inhibiting these enzymes, alkylresorcinols can delay the digestion and absorption of carbohydrates, resulting in a slower and more controlled release of glucose into the bloodstream.

In addition to their direct effects on glucose regulation, alkylresorcinols possess antioxidant and anti-inflammatory properties. Oxidative stress and chronic low-grade inflammation are closely associated with the development and progression of diabetes. By reducing oxidative stress and suppressing inflammation, alkylresorcinols may contribute to improved glycemic control and prevention of diabetes-related complications (Zhao, 2020).

It is important to note that although the studies conducted so far have provided promising evidence of the anti-diabetic activity of alkylresorcinols, further research is still needed to fully understand their mechanisms of action and evaluate their efficacy and safety in human subjects. Additionally, the optimal dosage, bioavailability, and potential side effects of alkylresorcinols need to be determined through rigorous clinical trials. These actions could contribute to better glycemic control and may help in managing diabetes or preventing its onset.

6.8.2 Anti-obesity activity

In addition to their anti-diabetic properties, alkylresorcinols have been studied for their potential anti-obesity effects. Obesity is a complex health issue characterized by excessive accumulation of body fat, leading to a higher risk of various chronic diseases such as cardiovascular disorders, diabetes, and certain types of cancer. Addressing obesity has become a global health priority, prompting scientists to explore novel approaches for prevention and management. Alkylresorcinols have emerged as a promising candidate in the fight against obesity due to their intriguing biological activities (Ross, 2012).

Several studies have demonstrated the potential anti-obesity effects of alkylresorcinols. One of the primary mechanisms underlying their action is their ability to inhibit adipogenesis, which is the process of fat cell formation. Alkylresorcinols have been found to interfere with the differentiation of preadipocytes into mature adipocytes, thereby reducing fat storage. This inhibitory effect on adipogenesis is crucial in controlling body weight and preventing obesity. Moreover, alkylresorcinols exhibit lipolytic activity, meaning they can promote the breakdown of stored fats within adipocytes.

This process, known as lipolysis, releases fatty acids, which can then be utilized for energy production. By enhancing lipolysis, alkylresorcinols can contribute to the reduction of adipose tissue and overall body fat mass, thereby aiding in weight management.

Another noteworthy aspect of alkylresorcinols is their potential impact on appetite regulation. Obesity is often associated with dysregulated appetite control, leading to increased caloric intake and difficulty in maintaining a balanced diet. Alkylresorcinols have shown promising effects in modulating appetite-regulating hormones, such as ghrelin and leptin, which play key roles in hunger and satiety. By influencing these hormonal signals, alkylresorcinols may help regulate appetite and promote feelings of fullness, ultimately contributing to weight control.

Furthermore, alkylresorcinols possess antioxidant and anti-inflammatory properties, which can indirectly support their anti-obesity effects. Oxidative stress and chronic low-grade inflammation are commonly observed in obesity and can contribute to its progression. By scavenging free radicals and reducing inflammation, alkylresorcinols may mitigate the negative impact of these processes on adipose tissue, metabolic health, and weight management.

It is important to note that while the anti-obesity potential of alkylresorcinols has been demonstrated in various *in vitro* and animal studies, further research is needed to establish their efficacy and safety in human subjects. Clinical trials evaluating the effects of alkylresorcinols on body weight, body composition, and metabolic parameters are warranted to provide stronger evidence (Zhao, 2020).

Overall, alkylresorcinols exhibit promising bioactivity, particularly about their anti-obesity effects. Their ability to inhibit adipogenesis, enhance lipolysis, regulate appetite, and exert antioxidant and anti-inflammatory actions make them an intriguing subject of research in the field of obesity management. However, additional studies are required to validate their efficacy, safety, and potential applications in human populations. Continued exploration of alkylresorcinols may pave the way for the development of novel therapeutic strategies to combat obesity and its associated health risks (Ross, 2012).

6.8.3 Cardioprotective effects

Alkylresorcinols have also shown potential cardioprotective effects, indicating a positive impact on heart health. Lipid metabolism refers to the processes involved in the synthesis, breakdown, and utilization of fats in the body. Studies have suggested that alkylresorcinols can modulate lipid metabolism by reducing the levels of harmful cholesterol, such as low-density lipoprotein (LDL) cholesterol, and increasing the levels of beneficial cholesterol, such as high-density lipoprotein (HDL) cholesterol.

Alkylresorcinols have been found to inhibit the activity of the enzyme 3-hydroxy-3-methylglutaryl-coenzyme A reductase (HMG-CoA reductase), which is involved in the synthesis of cholesterol in the liver. By inhibiting this enzyme, alkylresorcinols may help reduce cholesterol production, leading to lower levels of LDL (low-density lipoprotein) cholesterol, commonly referred to as "bad" cholesterol. Elevated levels of LDL cholesterol are a major risk factor for cardiovascular diseases such as atherosclerosis and coronary artery disease.

Furthermore, alkylresorcinols have demonstrated antioxidant properties. Oxidative stress plays a crucial role in the development and progression of cardiovascular diseases. Antioxidants help counteract the damaging effects of reactive oxygen species, thus protecting the cardiovascular system. By scavenging free radicals and reducing oxidative stress, alkylresorcinols may contribute to the prevention of cardiovascular diseases (Zhao, 2020).

In addition to their direct effects on lipid metabolism and antioxidant properties, alkylresorcinols have shown the potential in enhancing the therapeutic effects of certain medications. For instance, some studies have suggested that alkylresorcinols can enhance the lipid-lowering effects of statin drugs, which are commonly prescribed to reduce cholesterol levels. The combination of alkylresorcinols with statins may lead to a more significant reduction in LDL cholesterol and improve overall cardiovascular outcomes.

Moreover, alkylresorcinols have been investigated for their antiplatelet and anticoagulant effects. Platelet aggregation and blood clotting play a crucial role in the development of cardiovascular events, such as heart attacks and strokes. Alkylresorcinols have shown potential in inhibiting platelet aggregation and reducing blood clot formation, which could have beneficial effects on cardiovascular health (Ross, 2012).

It is worth noting that while the bioactivity of alkylresorcinols appears promising, further research is needed to fully understand their mechanisms of action and to establish their efficacy and safety in the context of cardiovascular therapeutics. Additionally, the bioavailability and metabolism of alkylresorcinols in the human body require further investigation to optimize their therapeutic potential.

Overall, alkylresorcinols exhibit bioactivity that may positively impact cardiovascular health, particularly through their effects on lipid metabolism and potential enhancement of therapeutic interventions. Their ability to inhibit cholesterol synthesis, act as antioxidants, and potentially improve the efficacy of lipid-lowering drugs make them an intriguing area of research for the prevention and treatment of cardiovascular diseases.

The multifaceted therapeutic potential of alkylresorcinols arises from their ability to target different physiological pathways. By regulating glucose metabolism, alkylresorcinols may contribute to better glycemic control in diabetes. Their inhibitory effects on adipogenesis make them potential candidates for managing obesity. Furthermore, their ability to modulate lipid metabolism and reduce cholesterol levels suggests a possible role in promoting cardiovascular health. However, it is important to note that the full extent of these beneficial effects and their underlying mechanisms is still being explored, and further research is needed to fully understand and harness the therapeutic potential of alkylresorcinols.

In conclusion, overall, alkylresorcinols demonstrate a wide range of bioactivities that can potentially benefit human health. Alkylresorcinols are natural compounds found in cereal products like wheat and rye. They have diverse biological activities, including antimicrobial, anti-inflammatory, antioxidant, and anti-cancer effects.

The antimicrobial activity of alkylresorcinols has been well-studied, and they have been shown to inhibit the growth of bacteria, fungi, and viruses. They disrupt microbial cell membranes, leading to cell death. This makes them potential natural antimicrobial agents for food preservation and pharmaceutical formulations (Zhao, R. 2020).

Alkylresorcinols also exhibit significant anti-inflammatory and antioxidant effects. They suppress the production of pro-inflammatory mediators and enzymes and act as free radical scavengers to protect cells from oxidative damage. Oxidative damage caused by ROS and RNS is linked to various diseases, and dietary antioxidants like alkylresorcinols can counteract these effects.

Alkylresorcinols have prebiotic activity, as they can stimulate the growth of beneficial bacteria in the gut, such as Lactobacilli and Bifidobacteria. They are also potential candidates for cancer treatment, as they induce apoptosis in cancer cells, inhibit angiogenesis, and possess antioxidant properties. Additionally, alkylresorcinols have neuroprotective effects, acting as antioxidants and anti-inflammatory agents to mitigate oxidative stress and inflammation in neurodegenerative diseases (Ross, 2012).

Briefly, alkylresorcinols are bioactive compounds with diverse therapeutic potential. However, further research is still needed to elucidate their mechanisms of action, optimize their bioavailability, and evaluate their long-term effects in humans. Harnessing the potential of alkylresorcinols may lead to the development of novel natural compounds or drugs for the prevention and treatment of various diseases (Zhao, 2020).

Chapter

7 OPTIMAL UTILIZATION OF BIOACTIVE MOLECULES IN THE MEDICAL FIELD

CONTENTS

7.1. INTRODUCTION

The most effective application of bioactive molecules in medicine has enormous potential to advance medicine and improve patient outcomes. Bioactive molecules are substances that have specific biological properties and can interact with cellular targets to have therapeutic effects. They can be naturally occurring or created. These molecules can be derived from a wide range of sources, including plants, animals, microorganisms, and laboratory-created compounds (Lewandowicz, 2022)

Bioactive molecules have numerous potential applications in the medical industry. They have shown significant efficacy in the treatment and prevention of disease due to their ability to specifically target molecular pathways involved in pathological processes. Bioactive molecules offer numerous therapeutic options, ranging from cancer therapy to neurological disorders, cardiovascular health, and metabolic disorders.

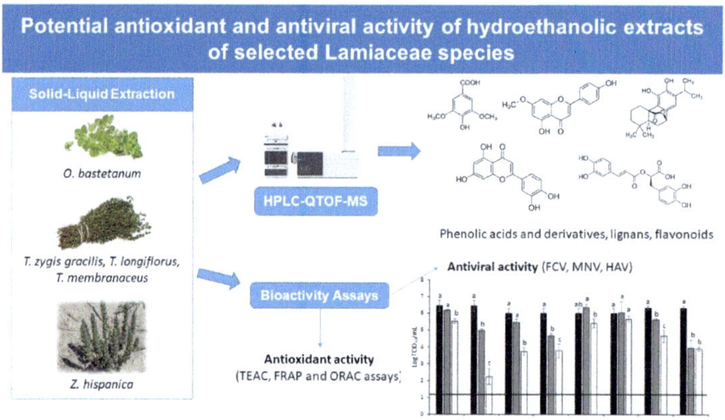

Figure 7.1. An illustration of bioactive molecules on viruses.

Source: Duque-Soto C, Borrás-Linares I, Quirantes-Piné R, Falcó I, Sánchez G, Segura-Carretero A, Lozano-Sánchez J. Potential Antioxidant and Antiviral Activities of Hydroethanolic Extracts of Selected Lamiaceae Species. *Foods*. 2022; 11(13):1862. https://doi.org/10.3390/foods11131862 [accessed 9 Jun, 2023]

Furthermore, advances in formulation and delivery methods have facilitated targeted drug delivery, ensuring that bioactive molecules reach the intended site of action with greater efficacy and fewer side effects. The ability to tailor bioactive molecules to an individual patient's needs and genetic makeup opens up new avenues for personalized medicine and precision therapy.

However, to fully utilize bioactive molecules, pharmacokinetics, bioavailability, safety, and regulatory requirements must all be addressed. Extensive research and development efforts are required to improve our understanding of these molecules, improve their formulation, determine their toxicity, and navigate the regulatory environment (Malcata, 2017).

7.2 BIOACTIVE MOLECULES IN REGENERATIVE MEDICINE AND TISSUE ENGINEERING

Tissue engineering and regenerative medicine are two areas of medical research that are showing promise in terms of repairing or replacing damaged organs and tissues. The use of bioactive molecules is one of the techniques used by these cutting-edge disciplines to promote tissue regeneration. Bioactive molecules are required to promote tissue growth, alter cellular behavior, and improve the integration of engineered constructs.

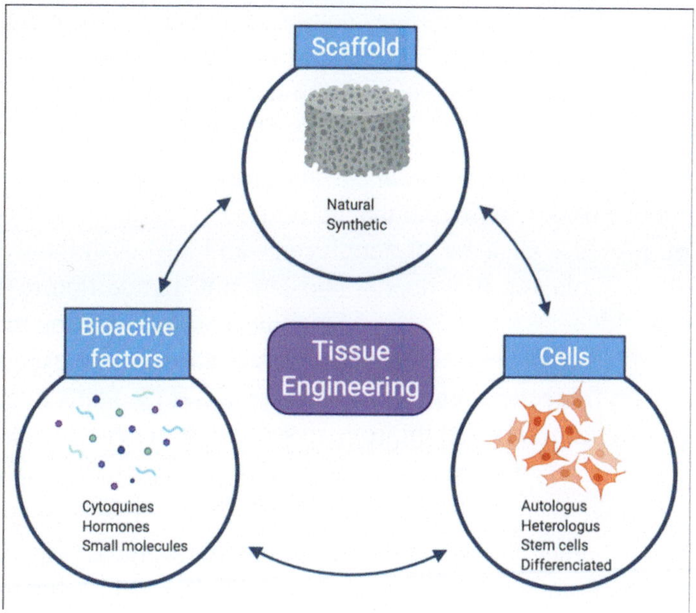

Figure 7.2. The tissue engineering triad.

Source: Doblado LR, Martínez-Ramos C and Pradas MM (2021) Biomaterials for Neural Tissue Engineering. Front. Nanotechnol. 3:643507. doi: 10.3389/fnano.2021.643507

Bioactive molecules and cellular signaling: Bioactive molecules are naturally occurring or artificial substances that interact with biological systems, particularly cells, and tissues. These molecules can include growth factors, cytokines, chemokines, extracellular matrix (ECM) components, peptides, and small molecules. They affect cells via cellular signaling pathways by activating intricate cascades of events that control cell behavior such as proliferation, migration, differentiation, and survival.

Cytokines and Growth Factors: Growth factors are essential for tissue growth, repair, and regeneration. Transforming growth factor-beta (TGF-β), vascular endothelial growth factor (VEGF), and fibroblast growth factor (FGF) are a few examples. They help with ECM synthesis and remodeling, as well as controlling cell migration, differentiation, and proliferation. These molecules are frequently added to biomaterial scaffolds or administered via controlled release systems to promote tissue regeneration.

Cytokines, on the other hand, are small proteins that play a role in cell signaling and communication. They are in charge of inflammatory processes, immune responses, and tissue repair. Interleukins, tumor necrosis factor (TNF), and interferons are cytokines that are essential in tissue engineering and regenerative medicine. By decreasing the inflammatory environment, cytokines can change the behavior of immune cells, promote angiogenesis, and improve tissue healing (Lewandowicz, G. 2022).

Extracellular matrix components: The ECM provides structural stability as well as biochemical cues for cellular activities. It is composed of a complex network of proteins such as proteoglycans and glycosaminoglycans, as well as collagen, elastin, fibronectin, and laminin. In addition to acting as physical scaffolds for cells, these ECM elements act as signaling molecules, controlling cell adhesion, migration, and differentiation. In tissue engineering, biomaterials are frequently functionalized with ECM elements to mimic the natural microenvironment and improve tissue integration and regeneration.

Small Molecules and Peptides: Peptides and small molecules have special advantages in regenerative medicine and tissue engineering due to their ability to interact with specific cellular receptors and activate signaling pathways. Peptide sequences derived from ECM proteins, such as RGD (Arg-Gly-Asp), can aid in cell adhesion and migration. Furthermore, small molecules such as receptor modulators, enzyme inhibitors, growth factor agonists, and antagonists have shown promise in regulating cellular activity and tissue regeneration. Because of their fine control over cellular processes, they are useful tools for tissue engineering applications (Malcata, 2017).

Tissue Engineering and Regenerative Medicine Applications: Bioactive molecules have a wide range of applications in tissue engineering and regenerative medicine. They've helped with organ transplants, bone regeneration, nerve regeneration, cardiovascular tissue engineering, and wound healing. By adding growth factors, cytokines, or ECM components to biomaterial scaffolds, researchers can create an environment that promotes cellular activity and tissue formation. Furthermore, bioactive molecules can be used to direct stem cell differentiation into specific cell lineages, assisting in tissue regeneration.

Future Obstacles and Directions: As the field of regenerative medicine advances, so will the use of bioactive molecules. Researchers are investigating cutting-edge delivery techniques such as hydrogels and nanoparticles to improve the controlled release and targeted delivery of bioactive molecules. Furthermore, research is ongoing to develop novel bioactive molecules with improved efficacy and stability.

However, several obstacles remain. It is critical to optimize the dosing, release kinetics, and spatial distribution of bioactive molecules within tissue engineering constructs. Regulatory hurdles and high manufacturing costs must be overcome to translate bioactive molecules into clinical applications.

7.3 BIOACTIVE MOLECULES FOR CANCER THERAPY

Cancer remains a major global health issue in terms of morbidity and mortality. Traditional therapies such as surgery, chemotherapy, and radiation therapy have long been used to treat cancer. However, these treatments frequently have serious drawbacks. As a result of the search for new and potent therapeutics, there has been an increase in the study of bioactive molecules derived from nature in recent years. Because of their diverse and complex chemical structures, these molecules offer promising opportunities for targeted cancer therapy.

Bioactive Molecules and cancer: Bioactive molecules are chemicals derived from living organisms such as plants, animals, and microorganisms. These molecules possess distinct biological properties and are required for a variety of physiological processes. When used in cancer therapy, bioactive molecules have several anti-cancer properties, including the ability to induce apoptosis, inhibit angiogenesis, and prevent metastasis. They can specifically target cancer cells or alter signaling pathways that contribute to tumor development and spread. Because of their various mechanisms of action, they are desirable candidates for the development of novel therapies that go beyond the limitations of traditional treatments.

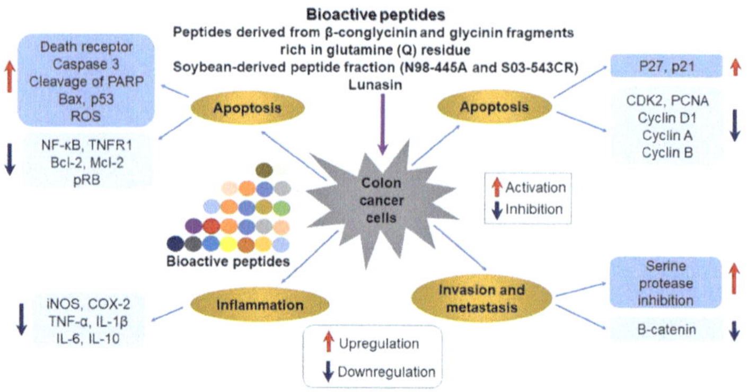

Figure 7.3. Soybean-Derived Bioactive Peptides the regulation of cancer.

Source: Kim I-S, Yang W-S, Kim C-H. Beneficial Effects of Soybean-Derived Bioactive Peptides. *International Journal of Molecular Sciences*. 2021; 22(16):8570. https://doi.org/10.3390/ijms22168570 [accessed 9 Jun, 2023]

There are several bioactive compounds originating from plants, and many of these chemicals show promise against cancer. . Taxanes produced by the Pacific yew tree (*Taxus brevifolia*), for example, have transformed the treatment of breast, lung, and ovarian cancers. These substances prevent cell division and induce apoptosis by interfering with microtubule dynamics. Similar to this, vincristine and vinblastine from the Madagascar periwinkle (*Catharanthus roseus*) have shown efficacy against leukemia and lymphoma by interfering with mitotic spindle formation (Malcata, . 2017).

Another notable example is curcumin, which is derived from the turmeric plant (*Curcuma longa*). Curcumin exerts strong anti-inflammatory and anti-cancer effects via a variety of mechanisms, including the modulation of various signaling pathways involved in tumor growth, apoptosis, and metastasis. Several clinical trials on various cancer types are being conducted to investigate its therapeutic potential.

Marine Molecules with Bioactive Properties: The Earth's oceans are a rich source of bioactive molecules with high potential for cancer treatment. Many anti-cancer substances are produced by marine organisms such as sponges, corals, and algae. Eribulin, a substance derived from marine sponges, is one such substance that has been approved for the treatment of metastatic breast cancer. Eribulin inhibits microtubule dynamics, resulting in apoptosis and cell cycle disruption (Lewandowicz, 2022).

Another notable example is trabectedin, which is derived from the marine organism *Ecteinascidia turbinate*. Trabectedin works against ovarian cancer and soft tissue sarcomas by inhibiting transcription factors by attaching to the DNA minor groove. Furthermore, dolastatins and cyanobacterial compounds, which are marine-derived peptides that target different cellular pathways, have shown promising anti-cancer activity.

Microbe-derived bioactive molecules: Microorganisms such as bacteria, fungi, and actinomycetes produce a large number of bioactive molecules with anti-cancer properties. Doxorubicin and bleomycin, two Streptomyces-derived drugs, are commonly used in chemotherapy. These substances inhibit cancer cell DNA replication and cause DNA damage, ultimately killing the cells. Epothilones and enediynes are two more microbial compounds that have shown promise as anti-cancer drugs.

Future challenges and opportunities: Despite the immense potential of bioactive molecules in cancer treatment, their development and clinical translation face numerous challenges. Obtaining enough of these molecules from natural sources can be difficult and costly. Furthermore, because of their complex chemical structures, the synthesis, modification, and formulation of these compounds for the best drug delivery are frequently complicated. Rigid preclinical and clinical evaluations are also required to determine their safety, effectiveness, and pharmacokinetic profiles.

Nonetheless, technological advances such as synthetic biology, combinatorial chemistry, and high-throughput screening are assisting in the development of bioactive molecules. Furthermore, by combining computational and bioinformatics methods, it is possible to predict drug-likeness, identify potential targets, and accelerate the drug discovery process. Bioactive molecules derived from natural sources have enormous potential for the development of efficient and specific cancer therapies. Because of their complex chemical structures and diverse mechanisms of action, these molecules are invaluable in fighting cancer and overcoming the limitations of conventional therapies. The potency of nature's pharmacological repertoire in the form of plant-derived, marine-derived, and microbial bioactive molecules has the potential to improve outcomes and quality of life for cancer patients (Malcata, 2017).

7.4 BIOACTIVE MOLECULES IN NEUROLOGICAL DISORDERS

A wide range of conditions known as neurological disorders that affect both the central and peripheral nervous systems cause various functional impairments.

These diseases, such as epilepsy, multiple sclerosis, Alzheimer's disease, and Parkinson's disease, pose serious public health issues worldwide. In recent years, there has been a growing interest in the potential of bioactive molecules as therapeutic interventions for neurological disorders. Bioactive molecules, whether created in a lab or derived from nature, have unique pharmacological properties that can alter neural pathways and open up therapeutic options (Lewandowicz, 2022).

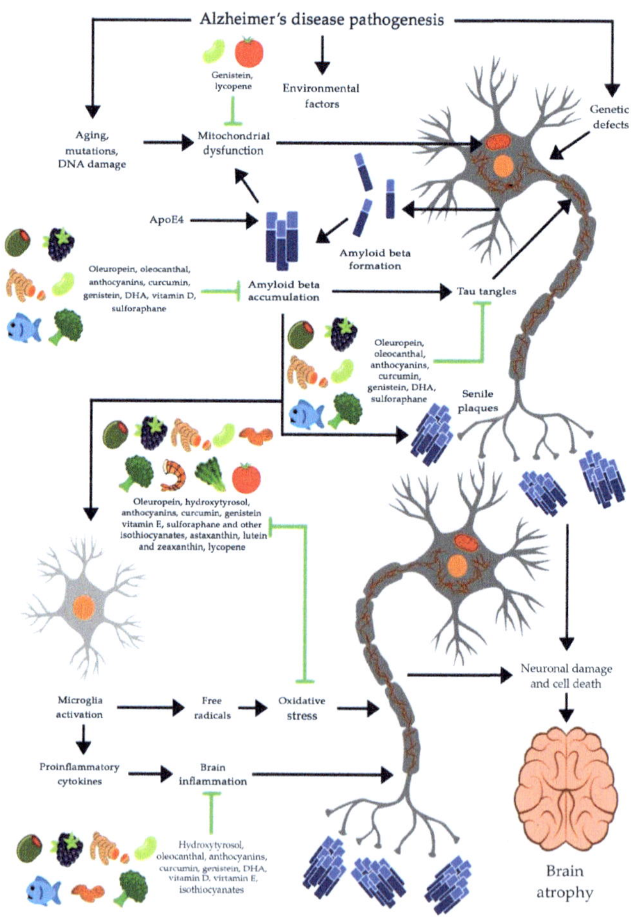

Figure 7.4. Action of bioactive compounds on Alzheimer's disease.

Source: Grodzicki W, Dziendzikowska K. The Role of Selected Bioactive Compounds in the Prevention of Alzheimer's Disease. *Antioxidants*. 2020; 9(3):229. https://doi.org/10.3390/antiox9030229 [accessed 9 Jun, 2023]

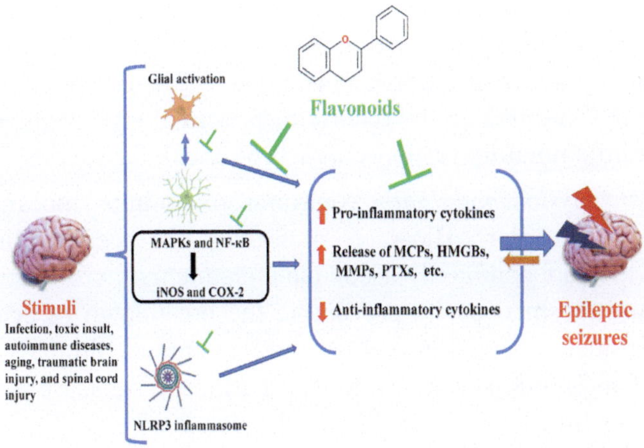

Figure 7.5: A summarized illustration of the potential pharmacological effects of bioactive compounds on epilepsy.

Source: Rabidas SS, Prakash C, Tyagi J, Suryavanshi J, Kumar P, Bhattacharya J, Sharma D. A Comprehensive Review on Anti-Inflammatory Response of Flavonoids in Experimentally-Induced Epileptic Seizures. *Brain Sciences*. 2023; 13(1):102. https://doi.org/10.3390/brainsci13010102 [accessed 9 Jun, 2023]

The biological mechanisms of bioactive chemicals: Bioactive molecules are organic or inorganic chemical substances that interact with biological systems and have a specific impact on cellular functions. In the context of neurological disorders, bioactive molecules exhibit a variety of mechanisms of action, including neuroprotection, anti-inflammatory properties, neurotransmission modulation, antioxidant effects, and neuronal regeneration stimulation. These mechanisms are critical in preventing the pathophysiology that underpins neurological disorders as well as promoting functional recovery.

Neuroprotection effects: Neurological disorders are frequently associated with neuronal loss or damage. Flavonoids, polyphenols, and cannabinoids are examples of bioactive compounds with neuroprotective properties because they prevent neuronal cell death and maintain brain function. For example, polyphenol resveratrol, found in red grapes and berries, has been shown to reduce oxidative stress and inflammation while increasing neuronal survival in neurodegenerative disorders.

Anti-inflammatory properties: Neurological disorders are exacerbated in part by inflammation. Both omega-3 fatty acids, which are abundant in fish oil, and curcumin, a bioactive molecule found in turmeric, have potent anti-inflammatory properties. By modulating several inflammatory pathways, including the nuclear factor-kappa B (NF-B) signaling pathway, these molecules can reduce the release of pro-inflammatory mediators and improve the inflammatory response in neurological disorders.

Neurotransmission modulation: Neurotransmitter imbalances are common in neurological disorders. Bioactive molecules like acetylcholine esterase inhibitors, which are commonly used in the treatment of Alzheimer's disease, improve neurotransmission by preventing the breakdown of acetylcholine, a neurotransmitter involved in memory and cognition. Similarly, dopamine agonists such as levodopa are used to treat Parkinson's disease symptoms by increasing dopamine levels in the brain (Malcata, 2017).

Antioxidant effects: Neurological disorders are characterized by oxidative stress, which is defined as an imbalance between the production of reactive oxygen species (ROS) and the antioxidant defense system. Bioactive molecules with antioxidant properties, such as vitamin E, flavonoids, and coenzyme Q10, scavenge free radicals and protect neurons from oxidative damage, slowing disease progression and improving neurological outcomes.

Neuronal regeneration and plasticity: In neurological disorders, ability to regenerate and form new connections is frequently compromised. Bioactive molecules such as brain-derived neurotrophic factor (BDNF), nerve growth factor (NGF), and specific herbal extracts promote neuronal regeneration and synaptic plasticity. These molecules promote the formation of new neurons, the survival of existing neurons, and the formation of useful neural networks, thereby restoring cognitive and motor functions (Lewandowicz, 2022).

7.5 TARGETED DRUG DELIVERY USING BIOACTIVE MOLECULES

Drug delivery systems are thought to improve pharmaceutical efficacy and therapeutic index by shielding and stabilizing bioactive molecules, such as proteins and peptides, from body fluid enzymes and/or inappropriate physicochemical conditions, while protecting healthy tissues from toxicity. When administered intravenously or topically, liposomes are biocompatible, and biodegradable, and do not elicit an immune response. The ability to load any drug or complex molecule, hydrophobic or hydrophilic, continues to be their most important feature.

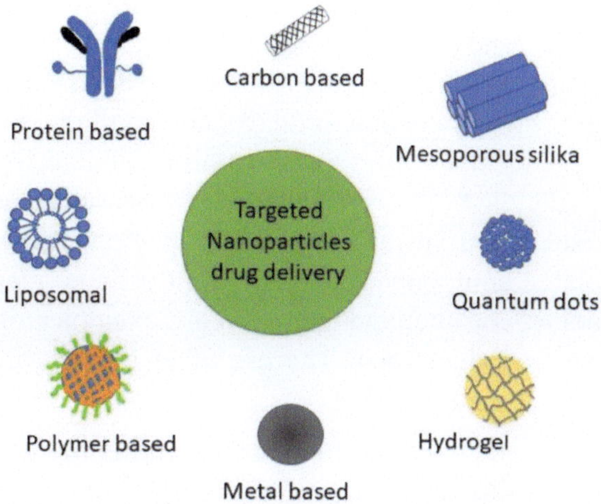

Figure 7.6. Nanotechnology and targeted drug delivery.

Source: Herdiana Y, Wathoni N, Shamsuddin S, Joni IM, Muchtaridi M. Chitosan-Based Nanoparticles of Targeted Drug Delivery System in Breast Cancer Treatment. *Polymers*. 2021; 13(11):1717. https://doi.org/10.3390/polym13111717 [accessed 9 Jun, 2023]

It is critical to select the appropriate lipid components, ratios, and thermosensitivity when developing an effective nano-liposomal formulation. Nano-liposomal surfaces can interact with target cells effectively, avoiding unfavorable interactions with plasma proteins and extending their half-life in the bloodstream. Liposome internalization and trafficking within target cells are mediated by mechanisms other than clathrin and caveolae, cell membrane cholesterol-dependent processes, and macropinocytosis-dynamin-independent processes that deliver drugs that modulate cell function.

Nature and its contribution to the discovery of potential pharmacological compounds

Nature has long been recognized as the primary source of bioactive molecules of varying structural complexity. For thousands of years, these have been used to treat a wide range of physiopathological conditions, including wounds, infectious diseases, and other disorders. They are frequently used in crude extracts as component mixtures. Conventional drugs, on the other hand, are typically composed of synthetic, small molecules ranging in size from 900 to 1500 Da that were chemically synthesized to match a preselected target via high-throughput screening. Natural macromolecules such as proteins, peptides, and polysaccharides are being sought by researchers as part of an increasing number of studies on natural therapeutic biomolecules. Only 2% to 5% of human genome products are small therapeutic molecule targets. In the nineteenth century, anti-malarial quinine, opiate analgesic morphine, and salicylic acid were among the first active principles to be identified and isolated from natural extracts, paving the way for the use of naturally isolated compounds in medicine. Caffeine, nicotine, codeine, atropine, colchicine, cocaine, and capsaicin are examples of alkaloids that have been discovered since then. The use of unaltered natural molecules increased by 43% during the 1930s.

Despite an increase in published data on natural products and bioactive molecules following this period, fewer unmodified natural products entered clinical settings. Unaltered natural molecules have been supplanted in the last thirty years by semisynthetic and synthetic derivatives that were created chemically but were inspired by natural molecules. The primary drivers of this profile shift were modern high-throughput screening platforms and synthetic combinatorial approaches aimed at the rapid discovery of novel drug candidates analogous to or similar to natural molecules exhibiting special activities identified over time, conducted via semi-synthetic modifications or total synthesis, but still retaining their pharmacophore. Unlike the time-consuming and costly methods of identifying natural compounds in crude extracts, followed by isolation and bulk acquisition, the generation of synthetic/semisynthetic analogs has a positive impact on the development of new, unaltered natural pharmacological compounds for clinical use. To maximize pharmaceutical production, the chemical synthesis of natural products was prioritized. This resulted in higher yields, higher quality, and purer bioactive compounds at a lower cost. Salicylic acid, the first natural product, was created in 1853 by imitating a natural molecule.

As a result of the cumulative knowledge gained from studies on bioactive molecules over the past two decades, the number of FDA-approved compounds classified as natural-based synthetic, semi-synthetic, and purely natural compounds increased from 34% of the top 100 best-selling medicines in 2010 to approximately 50% today (Lewandowicz, 2022).

Natural bioactive substances are found in both edible and non-edible organisms. Plants such as fruits and vegetables, whole grains, legumes, and oils are examples, as are animals, and microorganisms such as bacteria and fungi, algae, and other living things. Most research has been conducted on polyphenols such as curcumin, resveratrol, epigallocatechin gallate (EGCG), quercetin, and anthocyanin, followed by polyunsaturated fatty acids such as docosahexaenoic acid (DHA), eicosapentaenoic acid (EPA), and arachidonic acid (ARA). The next generation of pharmacological compounds, on the other hand, is expected to be composed of native proteins or enzymes and their bioactive peptides hidden within their primary structures. Milk proteins and plant lectins have been studied as non-toxic and highly effective treatments for a wide range of diseases. Individuals' health can be improved by bioactive polysaccharides such as lentinan and groan, vitamins, carotenoids, alkaloids, and even mineral elements such as Zn, Fe, Mg, Ca, Na, and P. These natural substances have numerous health-promoting properties, such as neuroprotective, antiviral, antitumor, immunomodulatory, anti-diabetic, and anti-obesity activity.

As a result, natural bioactive compounds and their structural analogs called "naturally inspired molecules, have had a significant impact on drug discovery history, particularly in the development of cancer and infectious disease treatments. These molecules continue to contribute to the development of new active agents in natural, synthetic, and semisynthetic forms that can treat the vast majority of medical conditions (Malcata, 2017).

Drug delivery system development and nanotechnology in clinical therapeutics

The method of administration and rate of release is primarily responsible for ensuring a sustained drug amount in the bloodstream that corresponds to the therapeutic index between the maximum safe concentration and above the minimum effective concentration, thereby avoiding abrupt peak concentrations due to massive drug release or the need for multiple doses to achieve drug effectiveness. Researchers were looking for an ideal drug delivery system to replace pills and capsules that release active ingredients in aqueous media, also known as an immediate release (IR) formulation. Spansules®, a first-generation drug delivery system made up of micro-pellets coated in varying thicknesses of water-soluble wax, allowed for the controlled release of oral medications for up to 12 hours while maintaining a constant blood concentration. During the first-generation drug delivery period, which lasted from the 1950s to the 1970s, transdermal formulations based on controlled release, primarily through dissolution and diffusion, were developed. Numerous issues with both delivery systems prompted the development of next-generation systems.

From 1980 to 2010, the second generation of drug delivery formulations was developed with the understanding that maintaining constant drug concentrations was not necessary. Intelligent polymers and hydrogels, biodegradable microparticles, solid implants, and *in situ* gel-forming implants capable of sustained release, as well as stimuli-responsive bioactive compounds, emerged during this time. Nanoparticles made of biodegradable polymers, polymeric micelles, lipids, chitosan, and dendrimers were developed to carry anticancer agents and gene sequences. To overcome biological and physicochemical barriers, the third generation of drug delivery systems was developed, which use advanced nanomaterials to deliver drugs or cell components such as peptides, proteins, DNA, and RNA sequences. Novel drug delivery concepts include nanoparticle-targeted drug delivery and self-regulated drug delivery. Despite an abundance of experimental data demonstrating that nanotechnology-based drug delivery systems could improve the efficacy of anticancer drugs against tumors in animal models, the FDA has only approved a small number of drugs.

According to the National Nanotechnology Initiative (2021), nanotechnology is the manipulation of nanoparticles ranging in size from 1 to 100 nm. It is used in a variety of fields such as engineering, physics, and information technology. Nanotechnology is regarded as the most promising treatment for degenerative pathologies such as cancer and central nervous system disorders, as well as an antiviral therapy to aid in SARS-CoV-2 vaccination and treatment. Pharmaceutical applications of nanotechnology, on the other hand, may have the greatest impact on human health (Lewandowicz, 2022).

Because of the morphology of nano-scaled particles, nanomaterials can now access previously inaccessible physiological sites, such as specific regions of the brain damaged by synucleinopathies or brain tumors, which require the ability to cross the blood-brain barrier for therapeutic intervention success. Furthermore, due to their high surface-to-volume ratio, nanomaterials can encapsulate bioactive molecule nanocarriers via adsorption, core entrapment, or covalent surface binding. This improves the drug's effects on the target sites (cells, tissues, or organs). The bioactive molecule is nano-encapsulated and continuously released under controlled conditions to achieve pharmacological effects while minimizing side effects. This reduces the number of required drug dosages when compared to conventional pharmaceuticals (Malcata, 2017). Nanoparticle-encapsulated compounds are resistant to oxidation in the bloodstream. They can still enter intracellular compartments through passive permeability, releasing bioactive substances into the cytoplasm or directing them to intracellular targets with ligands linked to nano-capsule internalization. Anti-tumorigenic therapies can boost the passive accumulation of anticancer drugs at the tumor site by increasing the enhanced permeability and retention effect (EPR), a phenomenon caused by the local inflammatory status of tumor blood vessels, which become permeable, allowing the passage and accumulation of nanometric materials in tumor tissues.

The most common chemical components of nanoparticles used in medical therapy are organic or inorganic compounds. Lipid-based substances such as liposomes and micelles, as well as natural and synthetic polymers such as chitosan, collagen, glycerol, polylactic-co-glycolic acid (PLGA), and dendrimers, are examples of the former. Inorganic nanoparticles include carbon nanotubes, quantum dots, metal oxides, and gold nanoshells. Through adsorption or covalent bonding to the surface of the nanocarrier, these nanoparticles can encapsulate therapeutic molecules or form complexes with molecules. These nanoconjugates could be used to treat cancer and degenerative diseases like atherosclerosis and Parkinson's disease, depending on the bioactive compound they contain.

It has been shown that using hydrophobically modified chitosan nanoparticles to deliver silibinin, a flavo-lignan isolated from the seeds of the milk thistle plant, improves response and increases the solubility of this compound with poor water solubility. Similarly, when encapsulated in albumin nanoparticles (NPs) and functionalized with the tripeptide arginine-glycine-aspartate (RGD), resveratrol (RV), a polyphenol non-flavonoid commonly found in red or dark grapes, increased content surrounding a target tumor by 8.1% compared to free-RV. In a murine model, RV-free treatment resulted in progressive tumor growth, whereas RV-NPs-RGD suppressed tumor growth without relapse after 35 days of treatment, indicating that it is a promising chemotherapy agent.

Curcumin, a naturally occurring antioxidant and anticancer chemical derived from turmeric or saffron that has been thoroughly studied, has limited clinical use because of molecular instability and poor solubility. Unlike free curcumin, solid lipid nanoparticle-encapsulated curcumin has a powerful antiproliferative effect on *SKBR3* cancer cells, with nanoencapsulation increasing bioavailability. Furthermore, at high extension, nano-encapsulated curcumin inhibited cell migration and induced cell apoptosis in 36.7% cells in *SKBR3*. With an IC50 of 0.52 M, co-encapsulation of tocotrienol (TRF) and simvastatin (SIM) in lipid nanoparticles was found to have an anti-proliferative effect on the breast adenocarcinoma +SA cell line. For 48 hours, approximately 25 to 40% of SIM is released, with the fastest release occurring within the first ten hours. The kinetic release will be slower and more controlled over the next 38 hours. When lianol (LN) was encapsulated in solid nanoparticles, similar effects were observed in HepG2 hepatocarcinoma and A549 pulmonary adenocarcinoma cell lines.

In comparison to nanoparticles without vincristine loading, folic acid conjugated through ionic interactions with chitosan nanoparticles (FA-CS) induces apoptosis in 75% of NCI-H460 lung cancer cells. When prepared at a 4:25 (v/v) vincristine/FA-CS ratio, nanoencapsulation efficiency rises to 95% and loading capacity rises to 48.65%, bolstering the promising anticancer effects on tumorigenic lung cells. Metformin (MET) nanofibers coated with PLGA (poly lactic-co-glycolic acid) demonstrated cytotoxicity against A549 lung cancer cells 48 hours after treatment when compared to free MET (Lewandowicz, 2022).

Liposome encapsulation improves the pharmacological effects of bioactive molecules, including metallic complexes, by lowering effective doses. This aids in the reduction of toxicity. Encapsulating iridium III complexes (Ir-1, Ir-2, and Ir-3) in nano-liposomes improves their anticancer activity against HepG2, hepatocellular carcinoma, HTC-116, colon cancer, HeLa, cervical cancer, A549, lung carcinoma, BEL-7402, SGC-7902, and gastric Free Ir-2. Encapsulated complexes stimulate apoptotic activity *in vitro* by increasing intracellular ROS and arresting the G0/G1 cell cycle. TPGS-coated liposomes loaded with doxorubicin (Dox) and SiRNA-corona Bcl-2 significantly reduced tumor mass in mice by a factor of seven, similar to free Dox. Poly(amidoamine) (PAMAM) dendrimer nanoparticles efficiently delivered methotrexate and D-glucose while also inhibiting MDA-MB 231 breast cancer cells.

Nanoparticles can be used to treat Parkinson's disease and atherosclerosis in addition to anti-tumor treatment. LXR-L, the liver receptor ligand, has been delivered to macrophages associated with atherosclerotic plaques using mannose-functionalized dendrimer nanoparticles (mDNP). In LDL-receptor knockout mice, four weeks of mDNP-LXR-L administration reduced atherosclerotic plaque, necrosis area, and inflammatory response by 10%, as measured by the expression of the metalloproteinase-9 (MMP-9) gene, which is controlled by NF-kB, the nuclear factor kappa B. There were no changes in the expression of plasma lipids or hepatic lipogenic genes. Gold nanoparticles (AuNCs) containing N-isobutyryl-L-cysteine (L-NIBC) have shown promise in the treatment of Parkinson's disease. 1-methyl-4-phenyliridine (MPP) nanoparticles inhibited the formation of Lewy bodies, neuron death and dysfunction, and synuclein fiber aggregation and formation in the brain, all of which are histological features of Parkinson's disease. These findings are consistent with those obtained in mice pre-inoculated with the neurotoxin 1-methyl-4-phenyl-1,2,3,6-tetrahydropyridine (MPTP) and treated with AuNCs, which aided in the regression of behavioral abnormalities identified by an open field, swimming, and rotarod tests. This adds to the evidence that L-NIBC-loaded nanoparticles could be used to treat Parkinson's disease.

Nanoparticles have improved the therapeutic efficacy of several pharmaceuticals by demonstrating the high potential of drugs complex to nanomaterials to treat neoplasms, neurodegeneration, and atherosclerosis, among other physiopathological conditions. But excluding pre-clinical research and large-scale manufacturing, as well as phase 0, I, II, and III clinical trials that must come before submission and approval by health regulatory agencies to reach phase IV, the American Pharmaceutical Association estimates that the period needed for optimisation and regulation leading to commercialization can be at least thirteen years. (Malcata, 2017).

Doxil®, the first drug-functionalized nanoparticle formulation approved in the United States, is a doxorubicin (DOX) nano-liposomal formulation. The Food and Drug Administration (FDA) was established in 1995. Doxil® can accumulate in solid tumors due to the EPR phenomenon and is an effective treatment for a variety of cancers, including metastatic ovarian cancer and AIDS-related Kaposi's sarcoma.

Doxil®'s ability to interleave DNA and inhibit topoisomerase II, thereby reducing DNA replication and RNA transcription, has been linked to an antitumorigenic effect. Doxil® is not easily removed from the plasma by the mononuclear phagocytic system (MPS), allowing for continuous release of encapsulated DOX and, as a result, superior therapeutic performance over free-DOX. Another chemotherapeutic formulation, Abraxane®, encapsulates paclitaxel in albumin nanoparticles. Abraxane®, which was approved by the FDA for the treatment of breast cancer in 2005, was added to the list of available treatments for advanced pancreatic carcinoma in 2013. Because of the formulation's improved bioavailability, endothelial transcytosis mediated by the albumin receptor (gp60) allows for higher intra-tumoral paclitaxel concentrations (Lewandowicz, 2022).

Liposomal formulation performance: characteristics, functionalization, and internalization

It is possible to create an effective liposomal formulation by selecting the optimal lipid composition, performing the necessary functionalization, and employing a clever targeting strategy. The selection of phospholipids is critical for determining the safety, stability, and efficacy of liposomes, taking into account the head group, chain length, and quantity of liposome components. Liposome efficacy is affected by their rigidity, surface charge, and lipid organization, in addition to their chemical composition, because surface modifications affect liposome stability and can be manipulated to increase their utility.

Glycerophospholipids, which are amphiphilic lipids composed of two saturated or unsaturated fatty acid chains and a glycerol molecule linked to a phosphate group, are the primary components of liposomes. The most common components used to make liposomes are phosphatidylcholine and phosphatidylethanolamine, both of which are abundant in plants and animals. Liposomes can acquire positive, negative, or neutral charges depending on the phospholipid head and chain, which determines their overall properties and functions. Liposomes can be made more stable by using phospholipids with longer tails and low levels of unsaturation and ether linkages. Longer saturated hydrocarbon chains in phospholipids allow them to interact and form tightly ordered bilayer structures. Shorter unsaturated hydrocarbon chains, on the other hand, result in the formation of fluid and disordered bilayers in liposomes.

1,2-dipalmitoyl-sn-glycero-3-phosphocholine (DPPC), 1,2-distearoyl-sn-glycero-3-phosphocholine (DSPC), 1,2-dioleoyl-sn-glycero-3-phosphocholine (DOPC), and 1,2-distearoyl-sn-glycero-3-phosphoglycerol (DSPG) are synthetic phospholipids.

The hydrophilic contacts between polar head groups, van der Waals forces between hydrocarbon chains, which hold the long hydrocarbon tails together, and hydrogen bonding with water cause the liposomes to spontaneously self-assemble into a closed bilayer while repelling the hydrophobic chains. Liposomes are classified as neutral, anionic, or cationic based on the elements of the lipid bilayer, and they can take on various 3D structures as unilamellar vesicles (ULVs), multilamellar vesicles (MLVs), or multivesicular vesicles (MVVs). SUVs, LUVs, and GUVs are ULVs with a single-lipid-bilayer membrane with lengths ranging from 1 to 100 m. Unlike multilamellar vesicles, which have concentric layers, MVVs are made up of many smaller vesicles that encase the interior of a larger vesicle. A different liposome composed of two bilayer membranes, the double-layer vesicle, has a framework that can increase liposome stability and delay and sustain the release of their load.

Liposomes' stability can be improved by phospholipids, one of their primary constituents, cholesterol, glycol-derivatives such as propylene glycol and polyethylene glycol (PEG), and even polymers such as chitosan. These elements have potent effects on healthy tissues and cells, stimulating or inhibiting immune responses. When cholesterol is added to liposome bilayers, it changes their fluidity, decreases permeability, and improves their *in vitro* and *in vivo* stabilities. To stabilize the liposome membrane, cholesterol, a hydrophobic molecule, induces dense phospholipid packing and inhibits lipid chain interactions. The cholesterol molecule fits within the lipid bilayer because its aromatic ring is parallel to the fatty acid chains and its hydroxyl group is close to the hydrophilic region of phospholipids. Proteins such as albumin, transferrin, macroglobulin, and high-density lipoproteins (HDL) can interact with liposomes in the absence of cholesterol, destabilizing the liposomal membrane structure and decreasing the efficacy of the drug delivery system (Malcata, 2017).

Chitosan, for example, is a polymer that is used to modify the surface of liposomes, resulting in the formation of a protective shell on the liposome surface, primarily for oral drug delivery. Flexible lipid vesicles composed of glycols such as propylene glycol and polyethylene glycol (PEG) have been proposed to improve drug delivery systems for the skin. Liposomes' half-life in the bloodstream can be increased by different PEGs on their surfaces, ranging from a few minutes for stealth liposomes (also known as PEGylated liposomes) to several hours for conventional liposomes. Conventional liposomes are rapidly eliminated from the bloodstream after entering reticuloendothelial system organs and tissues such as the liver and spleen. This is one of their main disadvantages. PEG increases in the lifetimes of circulating liposomes are influenced by the amount of grafted PEG and the polymer's molecular weight. PEGylated liposomes containing PEG-1900 and PEG-5000 have longer bloodstream residence times than PEG-750 and PEG-120. The molecular weight determines the conformation of PEG polymers on the surface of liposomes as well as the density of PEG surfaces in the mushroom (low concentration) and brush (high concentration) regimes. Higher PEG concentrations (brush regimen) in liposomes increase phagocytosis resistance and decrease human complement system activation, whereas PEG concentrations between 5%

and 10% (molar ratio) improve liposome stealth. PEG-enhanced liposome surfaces, which resemble water-like structures and have a cloaking effect, provide a steric barrier that prevents protein adsorption to liposome surfaces and recognition by the mononuclear macrophage phagocytic system, which would otherwise result in rapid liposome clearance. When PEGylated liposomes are repeatedly administered at predetermined intervals into the veins of animals, they cause immune reactions that result in the loss of long-circulating characteristics and the acceleration of the blood clearance (ABC) phenomenon. After the first dose, the spleen is thought to produce anti-PEG IgM, which binds specifically to PEG chains when a second dose is given several days later. One of the primary opsonins, complement activation, is thought to increase the liver's ability to absorb subsequent doses. The incidence and severity of the ABC phenomenon are influenced by the dose and physicochemical properties of PEGylated liposomes, as well as the time between administration of the liposomes and the drug they contain. Several strategies for reducing the immunogenicity of the PEG moiety after multiple administrations have been investigated. PEG lipids with shorter alkyl chains, such as mPEG-DSPE and mPEG-CH, can detach more quickly from the lipid bilayer, dampening the ABC phenomenon (Lewandowicz, 2022).

7.6 PERSPECTIVES ON BIOACTIVE COMPOUNDS AND NANODELIVERY FOR CARDIOVASCULAR DISEASE TREATMENT

In the United States, cardiovascular disease (CVD) is the leading cause of death. It is most common in both developed and developing countries, and its prevalence is to blame for rising healthcare costs. Tobacco use, high blood pressure, high cholesterol, malnutrition, inactivity, and obesity are all risk factors for cardiovascular disease. The primary cause of coronary artery disease (CAD) is atherosclerosis, which results in the narrowing or occlusion of coronary arteries. Atherosclerosis is caused by plaques (fat and cholesterol deposits) in the arteries.

Cardiovascular (CV) disorders include endocarditis, congestive heart failure, irregular heartbeat, and congenital heart disease (present at birth). Between 2007 and 2017, the number of deaths from cardiovascular diseases increased by 21% globally due to population growth and aging; the majority of deaths occurred in low- and middle-income countriesThe World Health Organisation (WHO) estimates that 23,3 million deaths would be attributable to CVD by 2030. Tobacco use, dyslipidemia, hypertension, diabetes, and obesity are all risk factors for cardiovascular disease (CVD). These can be avoided by implementing preventative measures. A healthier diet is critical for preventing cardiovascular disease and moving toward an illness-free future (Malcata, 2017).

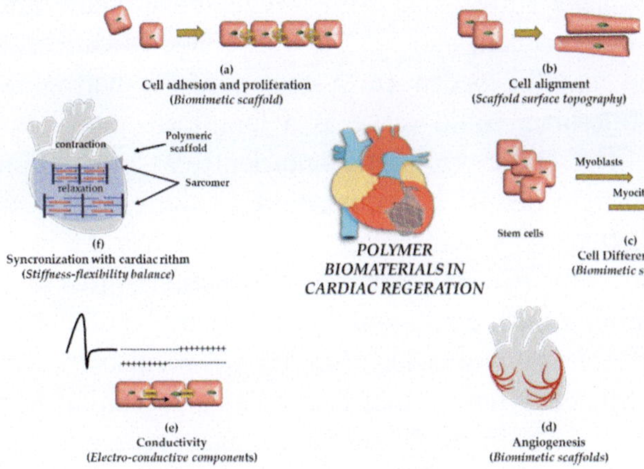

Figure 7.7. Biomaterials for cardiac applications.

Source: Trombino S, Curcio F, Cassano R, Curcio M, Cirillo G, Iemma F. Polymeric Biomaterials for the Treatment of Cardiac Post-Infarction Injuries. *Pharmaceutics*. 2021; 13(7):1038. https://doi.org/10.3390/pharmaceutics13071038 [accessed 9 Jun, 2023]

A poor diet is one of the leading causes of approximately 72% of all CVD deaths. Recent epidemiological studies show that plant-based diets can help prevent cancer and cardiovascular disease. The term "plant-based diet" refers to a variety of eating patterns that favor plant-based foods over animal products (such as meat and dairy). According to recent research, there is no agreement on how to define a "bioactive compound." However, it is widely accepted that these "complexes have the proficiency and aptitude with one or more parts of active tissue by predicting most of the effects." They have a positive effect on health and are being studied for the prevention of various diseases such as cancer, heart disease, diabetes, and so on. To name a few, bioactive substances include lycopene, resveratrol, lignan, tannins, and indoles. Bioactive compounds are classified based on their biochemical structure and functions. These compounds have been shown to have protective properties against certain immune-related pathologies, oxidative stress, and inflammation, in addition to being able to reduce LDL cholesterol oxidation and control endothelial nitric oxide amalgamation. Some of them are also low in estrogenic activity. These compounds may be able to delay the onset of age-related chronic diseases and regulate glucose metabolism.

Numerous epidemiological studies show a link between increased fruit and vegetable consumption and a lower incidence of cardiovascular disease. Numerous studies have shown that fruits, vegetables, whole grains, nuts, and seafood can help prevent and treat a wide range of cancers and heart conditions. Given that the ocean ecosystem contains half of all known species, aquatic microbes could serve as a long-term repository for rare bioactive metabolites. These oral dietary nutrients, when combined with a regular meal, will help people consume more of the elements thought to have therapeutic effects. Tea and coffee, two of the world's most popular beverages, are known to have numerous health benefits. Coffee seeds contain the caffeine-containing purine alkaloid 1,3,7-trimethylxanthine, which has several controversial latent properties. *Allium sativum*, also known as garlic, is a plant in the Alliaceae family that is used as a seasoning in Southeast Asia and Europe. It is high in flavonoids, organosulfur complexes, and other compounds that, when combined, provide a variety of health benefits. Environmental extraction techniques for obtaining bioactive components are becoming increasingly popular in the fields of dietary supplements and universal healthcare. Other pressing issues include extraction permits and pharmacological activities that occur during the extraction procedure. A growing number of bioactive substances are being studied in academia (Lewandowicz, 2022).

It can be difficult to consume all of the nutrients required for daily life to support a drug-based treatment for CVDs or to ensure that the body is functioning properly. As a result, nutraceutical encapsulation techniques have proven to be an effective strategy for preventing degradation caused by environmental factors such as temperature, light exposure, and UV exposure during production and storage. Furthermore, incorporating encapsulated nutraceuticals into various carriers addresses the primary issue of low bioavailability (as in the case of polyphenols), which is highly dependent on a variety of factors such as solubility, digestibility, absorption, and metabolism. Furthermore, the majority of bioactive substances are unstable in biological fluids' alkaline environments. By encapsulating the incorporated compounds in an appropriate matrix, they can be released at a specific concentration and time profile at a desired site of action. At the nanoscale, advanced nano-delivery systems improve the bioavailability and therapeutic efficacy of bioactive compounds. Effective nano-delivery systems, on the other hand, have the best properties for medication agent controlled release, long storage life, and enhanced therapeutic efficacy with no or minimal side effects.

The goal here is to keep you up to date on the key bioactive substances that are important in the prevention, diagnosis, and treatment of cardiovascular disease (CVD). By summarizing the most recent preclinical data and clinical trials reported in the scientific literature, describing the newer methodology and nanoformulation technologies that influence adequate drug or nutraceutical delivery, and outlining the most recent preclinical data, it is possible to discuss the current evidence demonstrating the advantages of nanoformulations and novel multifunctional nanomaterials capable of overcoming multiple obstacles, including physiology.

Bioactive compounds and their cardiovascular system therapeutic effect

The development of diseases such as atherosclerosis and cardiovascular disease is aided by oxidative stress. Various bioactive compounds' metabolic, anti-inflammatory, and antioxidant properties have been linked to their protection against atherosclerosis and cardiovascular disease (Malcata, 2017).

Carotenoids

Carotenoids are abundant in fruits and vegetables. They are classified into two groups based on their chemical structures: carotenes and xanthophylls. Carotenoids' antioxidant properties are beneficial to health. Carotenoids are found in the human body's organs and tissues. Carotenoids are abundant in tissues because they contain a large number of low-density lipoprotein receptors. Carotenoids can help prevent cardiovascular diseases like coronary heart disease and stroke. They have complex metabolic processes and are sensitive to external forces. As antioxidants, they increase catalase, glutamate dehydrogenase, superoxide dismutase, and glucan particles. Furthermore, carotenoids have been shown to inhibit IGF-1 activity and to be anti-cancer.

Anthocyanins are a type of polyphenol.

Berries contain a lot of micronutrients, fiber, and polyphenols, especially anthocyanins. Berries help to improve heart health. Anthocyanin derivatives have been extracted from cranberries, blueberries, and strawberries. Significant improvements were observed in LDL oxidation, lipid peroxidation, total plasma antioxidant potential, dyslipidemia, and glucose metabolism. It enhances endothelial function and plasma lipid profiles, activating endothelial nitric oxide synthase and lowering oxidative stress. As a result, abnormal platelet aggregation was reduced. Berries, which contain anthocyanins, are recommended as a key fruit category in a heart-healthy diet.

Anthocyanins, catechins, quercetin, rutin, and ellagitannins are all found in pomegranate juice. Pomegranate juice lowers blood pressure due to the antioxidant activity of anthocyanins and hydrolyzable tannins that act as radical scavengers (Malcata, 2017).

Lycopene

Lycopene, an oxygenated hydrocarbon carotenoid, has a structure that is very similar to that of -carotene. Lycopene has antioxidant properties due to the presence of conjugated double bonds. This fat-soluble pigment gives foods like tomatoes, guavas, and watermelons their distinctive red color. Cigarette smoking is a major risk factor for CVD. Smoking causes the body to produce free radicals, which are responsible for LDL oxidation, foam formation, and atherogenesis. Increased LDL oxidation is linked to atherosclerosis progression. Lycopene prevents coronary heart disease (CHD) by inhibiting LDL oxidation.

Lycopene and plasma levels in cardiovascular disease have been studied by researchers. The outcomes were scrutinized. Lycopene consumption was compared to the presence of estone. Lycopene was associated with a 17% reduction in CVD in this study. It works through a variety of mechanisms, including decreased biomolecule oxidation, antiangiogenic effects, lower cholesterol levels, apoptosis stimulation, and reduced inflammation (Lewandowicz, 2022).

Flavonoids

Flavonoids have been shown in numerous cohort studies and randomized controlled trials to reduce the risk of cardiovascular disease (CVD). Flavonoids improve LDL cholesterol, insulin sensitivity, and endothelial function. Flavonoids, specifically flavonols, anthocyanidins, proanthocyanidins, flavones, flavanones, and flavan-3-ols, have been shown in numerous studies to significantly reduce the risk of cardiovascular disease. The quality of flavonoid subcategories will be more important than the total flavonoid content in food. Furthermore, women have a stronger inverse relationship between flavonoid consumption and CVD risk than men. The evidence, however, is highly variable because randomized clinical trials are complementary.

Since its inception, flavonoid research has grown in popularity. One of the most difficult tasks was determining how to improve the bioavailability of flavonoids, both natural and synthetic. Flavonoids are in short supply, but this has recently been resolved. According to recent research, the gut and its microbiome play an important role in the synthesis of phenolic metabolites, which act as prebiotics and microbiota boosters.

Flavonoids have been shown to reduce gastrointestinal and gastric inflammation due to their ability to boost gut immune function. As a result, efforts to increase flavonoids' bioavailability focus primarily on improving their intestinal absorption. Boreol and methanol are common drug absorption enhancers. Methanol and borneol are both toxic, with methanol capable of causing blindness.

Regarding the CV System

In terms of antiatherosclerotic effects, the first step in the pathogenesis of atherosclerosis is the oxidative modification of low-density lipoproteins (LDL) by free radicals. Foam cells are formed when oxidatively modified LDL is rapidly absorbed by a scavenger receptor. Flavonoids are radicals that act as antioxidants by destroying chains. It has been demonstrated that quercetin and quercetin glycosides protect LDL from oxidative activation. Total plasma cholesterol levels and flavonoid intake have an inverse relationship, according to a Japanese study (Malcata, 2017).

Platelet aggregation is influenced by the composition of thrombotic disorders, which has anti-thrombogenic effects. Because activated platelets bind to the vascular endothelium and generate lipid peroxides and oxygen free radicals, prostacyclin and nitrous oxide (NO) cannot form in the endothelium. Tea pigment inhibits platelet adhesion and aggregation promotes fibrinolysis, and decreases blood clotting. Flavonoids like quercetin, kaempferol, and myricetin have been shown in animal studies to prevent platelet aggregation. Flavonols are particularly antithrombotic due to their ability to neutralize free radicals and keep endothelial prostacyclin and nitric oxide levels stable (Lewandowicz, 2022).

CVD is currently the leading cause of death in both advanced and developing economies in terms of cardioprotective effects. Atherosclerosis, coronary artery disease, arterial hypertension, and heart failure are all examples of cardiovascular system (CVS) disorders. Oxidative stress is the primary cause of CVS disorders. Endogenous oxidants and reactive oxygen/nitrogen species (RONS) are in equilibrium during oxidative stress, with free radicals dominating. Long-term flavonoid administration has been shown to reduce or be on the verge of reducing the incidence and consequences of cardiovascular disease. They excel at ferrous ion chelation, electron transfer, and the separation of reactive oxygen species. Over time, flavonoids may protect against the cardiotoxicity caused by cytostatic drugs like doxorubicin.

Flavonoids are vasorelaxants that lower arterial blood pressure and increase vasorelaxation. ED has a significant impact on the development of arteriosclerosis and arterial thrombosis, and it is a major contributor to the development of cardiovascular disease. Flavonoids help to prevent CVDs like atherosclerosis and hypertension. According to recent laboratory research, these polyphenols can lower arterial pressure and improve the vasodilation system. Flavonoids have long been known to cause an endothelial-dependent response. The anthocyanin delphinidin was also discovered to have an endothelium-dependent vaso-relaxing effect, according to the researchers.

To conclude, Various bioactive substances are being studied and evaluated to see if they have the potential to improve human health. They have antioxidant, anti-inflammatory, and anticarcinogenic properties that protect against infectious and metabolic diseases such as diabetes, cardiovascular disease, and cancer, in addition to physiological and cellular benefits. Their consumption has been linked to numerous health benefits, making them ideal ingredients for the development of new nutritional dietary supplements containing potent antioxidant and preservative properties.

Because of the global prevalence of CVDs, the high morbidity and mortality associated with these diseases, and the excessive side effects of current therapies, optimized alternative approaches for the prevention and/or treatment of these diseases are required. This article summarizes a large group of plant-derived bioactive compounds used as complementary therapies for cardiovascular diseases. To achieve these lofty goals, national and international campaigns promoting healthier, predominantly plant-based diets can be tailored to the target audience's personality and cultural preferences. The consequences of using bioactive substances to treat CVD are still being investigated. Some bioactive polyphenols and flavonols have been shown to lower risk factors for cardiovascular disease. Because of their antioxidant and anti-inflammatory properties, they are excellent candidates for the treatment of cardiovascular diseases. The current review, in addition to the detailed properties of bioactive compounds used in CVDs, provides an overview of the therapeutic benefits of nanoformulations and recent advances in this field (Malcata, 2017).

Although the efficacy of nanoformulations and theranostic nanoplatforms has been linked to the reported success of preclinical studies, the translational approach to clinical sectors has received little attention because the therapeutic applications of nano-phytomedicines in CVDs are still in the early stages of clinical development. Nanotechnologies will pave the way for more effective CVD treatment, a better prognosis, and fewer adverse side effects in non-target tissues. To effectively create and use low-cost nanoformulations, toxicological research, and additional research are required.

In conclusion, the most effective application of bioactive molecules has the potential to completely transform the medical industry. These molecules, which can be created in laboratories or from natural sources, have numerous therapeutic benefits. Using their natural properties, we can develop new treatments for a wide range of illnesses and conditions. Precision therapy, regenerative medicine, and targeted drug delivery systems are all promising ways to improve patient outcomes using bioactive molecules. Nonetheless, continuous research and development are required if they are to realize their full potential. Thorough safety evaluations, regulatory frameworks, and commercialization plans are required to ensure the efficient and ethical use of these molecules. With continued advancements and new trends, promising creative solutions, and better healthcare for all, the use of bioactive molecules in medicine has a promising future (Lewandowicz, 2022).

Chapter

8

BIOACTIVE LIPIDS

CONTENTS

8.1 INTRODUCTION

An intriguing class of substances known as bioactive lipids is involved in many vital biological processes. These many molecules play a variety of roles in cellular signaling, controlling inflammation, storing energy, and other physiological processes. Our understanding of human health and illness has been fundamentally altered by the research on bioactive lipids, which has produced ground-breaking findings. In this in-depth study, we explore the world of bioactive lipids, including their classification, uses, and effects on several facets of our existence.

Bioactive lipids come from a variety of lipid groups, including phospholipids, sphingolipids, glycolipids, and fatty acids. This class of compounds includes sphingosine-1-phosphate (S1P), ceramides, eicosanoids, endocannabinoids, lysophospholipids, and many others. These lipids are produced and degraded by intricate enzymatic processes, resulting in bioactive lipid mediators with specialized functions (Gambino, 2018)

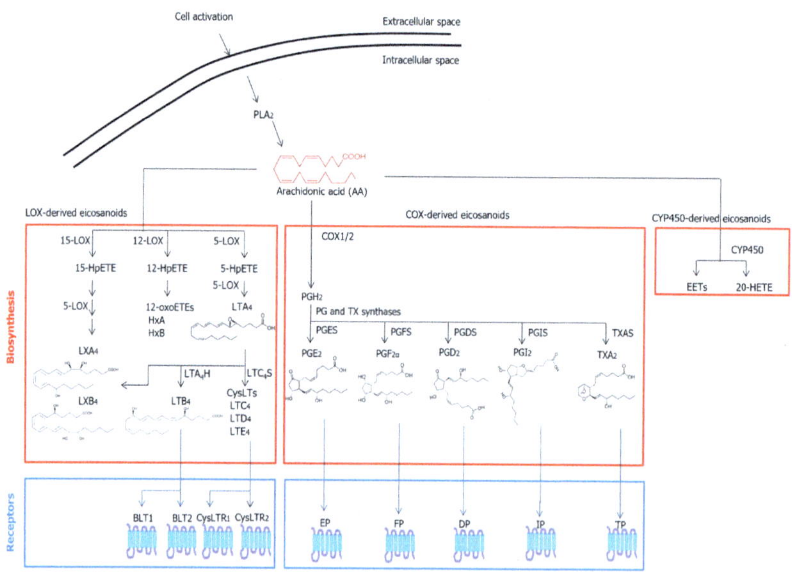

Figure 8.1. Metabolic pathways of bioactive lipids.

Source: Maccarrone M. Deciphering Complex Interactions in Bioactive Lipid Signaling. *Molecules*. 2023; 28(6):2622. https://doi.org/10.3390/molecules28062622 [accessed 6 Jun, 2023]

Ceramides, sphingosine-1-phosphate (S1P), and gangliosides are examples of the well-known class of bioactive lipids known as sphingolipids.

Ceramides, for instance, are involved in biological processes like cell division, death, and inflammation. Cell migration, immunity, and cell survival are just a few examples of the physiological activities that S1P regulates as a signaling molecule. Gangliosides, which are abundant in the nervous system and necessary for cell signaling, synaptic plasticity, and neuronal development.

Another significant class of bioactive lipids generated from arachidonic acid is called eicosanoids. Well-known eicosanoids implicated in inflammation, pain perception, blood clotting, and smooth muscle contraction include prostaglandins, thromboxanes, and leukotrienes. They serve as local messengers and have either paracrine or autocrine actions (Nicolaou, 2013).

Endocannabinoids are lipid mediators that interact with cannabinoid receptors in the endocannabinoid system, including anandamide and 2-arachidonoylglycerol (2-AG). These lipids control a number of physiological functions, such as appetite, mood, immunological response, and pain regulation.

Lysophospholipids play a role in cellular signaling, inflammation, and vascular homeostasis. Examples include lysophosphatidic acid (LPA) and lysophosphatidylcholine (LPC). They serve as ligands for particular G protein-coupled receptors and start intracellular signaling cascades when they are present.

Specific receptors and enzymes help the complex interaction of bioactive lipids in cellular signaling. Lipid mediators engage with cell surface receptors, setting off a series of actions that eventually control cellular reactions. Our understanding of bioactive lipid signaling has considerably benefited from the finding and characterization of these receptors and enzymes (Gambino, 2018). Recent years have seen a revolution in the study of bioactive lipids thanks to the development of cutting-edge analytical methods like mass spectrometry and lipidomics. These methods make it possible to identify, quantify, and profile different lipid species and their metabolites, giving us important information on how they affect both health and illness.

Understanding the roles and activities of bioactive lipids is essential for comprehending the complexities of human health and disease. Numerous clinical ailments, including cancer, neurological disorders, metabolic issues, and inflammation, have been connected to bioactive lipids. Furthermore, current studies are focusing on developing targeted therapies and discovering innovative medications, highlighting the huge therapeutic potential of these lipid molecules.

8.2 UNDERSTANDING BIOACTIVE LIPIDS

A wide range of lipid compounds known as "bioactive lipids" are essential components of several biological processes. These lipids have distinct chemical structures and have a significant impact on cellular signaling, the control of inflammation, and a number of physiological processes. Untangling the complexity of human health and disease requires an understanding of the relevance and functions of bioactive lipids (Nicolaou,2013).

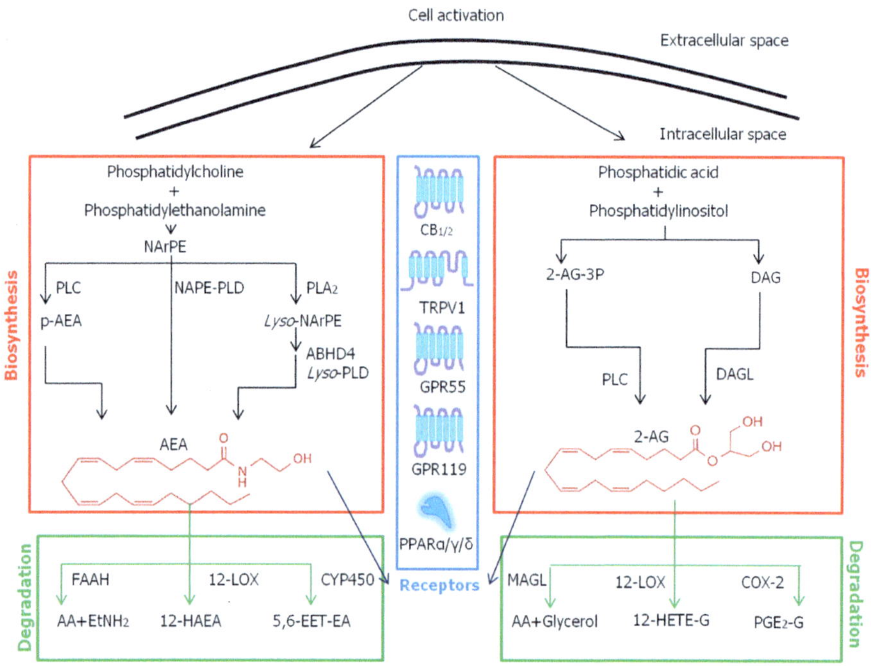

Figure 8.2. Receptors of bioactive lipids.

Source: Maccarrone M. Deciphering Complex Interactions in Bioactive Lipid Signaling. *Molecules*. 2023; 28(6):2622. https://doi.org/10.3390/molecules28062622 [accessed 6 Jun, 2023]

A large variety of chemicals originating from several lipid classes, including phospholipids, sphingolipids, glycolipids, and fatty acids, make up bioactive lipids. Sphingosine-1-phosphate (S1P), ceramides, eicosanoids, endocannabinoids, lysophospholipids, and many others are examples of bioactive lipids. Through complex enzymatic processes, these lipids are produced as bioactive lipid mediators, each of which has a distinct role. Our comprehension of cellular signaling pathways has been improved by the classification and study of bioactive lipids. These lipids serve as messengers, sending messages both within and between cells. As a result, they have an impact on vital procedures such as cell survival, proliferation, migration, and immunological responses. In numerous contexts, bioactive lipids have been thoroughly investigated. Sphingolipids, for example, have a role in cellular functions such as neural growth, inflammation, and apoptosis. Arachidonic acid-derived eicosanoids play significant roles in inflammation, pain perception, and blood coagulation. Endocannabinoids interact with cannabinoid receptors to modulate immunological response, hunger, and pain. Lysophospholipids function as signaling molecules and take part in physiological reactions like vascular homeostasis and inflammation (Gambino, 2018).

Understanding bioactive lipids has considerably benefited from developments in analytical techniques like mass spectrometry and lipidomics. These methods make it possible to identify, quantify, and profile different lipid species and their metabolites, giving us important information on how they affect both health and illness. Researchers want to create targeted therapies and therapeutic approaches to treat a variety of pathological ailments, such as inflammation, metabolic disorders, neurological diseases, and cancer, by understanding the roles and activities of bioactive lipids (Nicolaou, 2013).

8.3 CLASSIFICATION OF BIOACTIVE LIPIDS

A wide range of lipid compounds known as "bioactive lipids" have a substantial impact on cellular signaling, inflammation, and numerous physiological processes. Based on their chemical makeup, biosynthetic processes, and functional uses, these lipids can be divided into various groups. Understanding how bioactive lipids are categorized will help you better understand their unique characteristics and roles. Here, we investigate the principal categories of bioactive lipids:

- Sphingolipids: A class of bioactive lipids with a sphingoid base backbone are known as sphingolipids. They are essential for membrane shape, cell signaling, and the regulation of numerous cellular functions. Sphingolipids that are significant include:

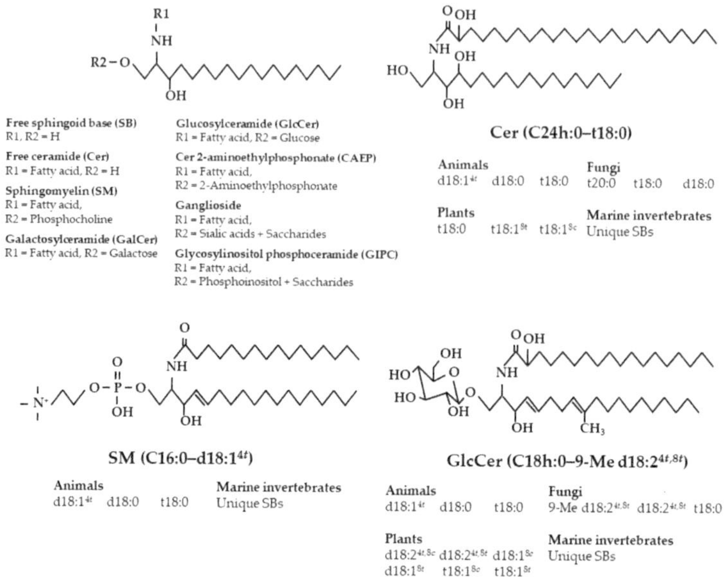

Figure 8.3. An image illustrating circulating sphingolipids.

Source: Yamashita S, Kinoshita M, Miyazawa T. Dietary Sphingolipids Contribute to Health via Intestinal Maintenance. *International Journal of Molecular Sciences*. 2021; 22(13):7052. https://doi.org/10.3390/ijms22137052 [accessed 6 Jun, 2023]

- Ceramides: A fatty acid is connected to a sphingosine backbone to form ceramides. They take part in biological activities such as apoptosis, cell division, and inflammation.
- Sphingosine-1-phosphate, often known as S1P, is a signaling molecule that controls a variety of physiological processes, such as cell migration, immunity, and vascular formation.
- Gangliosides: Found in large quantities in the neurological system, gangliosides are crucial for cell signaling, synaptic plasticity, and neuronal growth.

Eicosanoids: Polyunsaturated fatty acids like arachidonic acid are the source of eicosanoids, which are bioactive lipids. Inflammation, pain perception, blood clotting, and smooth muscle contraction all include them as local signaling molecules. Significant eicosanoids kinds include:

- Prostaglandins: Prostaglandins control vascular tone, platelet aggregation, and inflammation. They affect many physiological systems in a variety of ways.

- Thromboxanes: Thromboxanes aid in the production of blood clots by promoting platelet aggregation and vasoconstriction (Gambino, 2018).
- Leukotrienes, in particular those implicated in allergy and inflammatory reactions, have a role in the immunological response.

Endocannabinoids: In the endocannabinoid system, lipid mediators called endocannabinoids interact with cannabinoid receptors. They control a wide range of physiological activities, such as the control of appetite, mood, and immunological response. Endocannabinoids include, for instance:

- Anandamide is involved in the control of pain, mood management, and neuroprotection.
- 2-Arachidonoylglycerol (2-AG): 2-AG is involved in the control of hunger, neuroinflammation, and synaptic plasticity.

Lysophospholipids: When phospholipids are hydrolyzed, a single fatty acid chain is left behind, connected to a glycerol backbone. They act as signaling molecules and take part in cellular reactions like vascular homeostasis and inflammation. Examples worth mentioning are:

- Lysophosphatidic Acid (LPA) is a molecule that controls a variety of biological functions, such as cell division, migration, survival, and inflammation (Nicolaou, 2013).
- Lysophosphatidylcholine (LPC): LPC plays a role in cardiovascular disease, atherosclerosis, and inflammation.

Phospholipids: A significant class of lipids that forms the structural framework of cell membranes are phospholipids. Some phospholipids have bioactive qualities, while the majority are used as structural elements. For instance:

- Phosphatidic Acid (PA): PA is involved in intracellular signaling pathways and acts as a precursor for the creation of other bioactive lipids.
- Phosphatidylserine (PS): PS has a role in blood coagulation, apoptosis, and cell signaling.
- PI and its phosphorylated derivatives serve important roles in intracellular signaling, controlling things like cell proliferation, differentiation, and vesicular trafficking. PI is a kind of phosphatidylinositol.

Steroids: A four-ring structure distinguishes the class of lipids known as steroids. Some steroid hormones, though not all steroids are thought to be bioactive, show significant biological action. Examples are:

- Corticosteroids: Corticosteroids, like cortisol, control inflammation, affect immunological responses, and influence stress responses.
- Sex hormones: Steroid hormones such as estrogen, progesterone, and testosterone control sexual development, reproductive function, and other physiological functions.

Oxylipins: Polyunsaturated fatty acids are oxidized to produce oxylipins, which are bioactive lipids. They participate in cellular signaling, immunological responses, and inflammation. Oxylipins include a variety of substances, including:

- HETEs (hydroxyeicosatetraenoic acids) play a role in cell proliferation, vasoconstriction, and inflammation.
- Epoxyeicosatrienoic acids (EETs) and epoxydocosapentaenoic acids (EDPs) are two types of epoxides that play a role in the control of inflammation, blood pressure, and vascular tone.
- Resolvins and Protectins are specialized pro-resolving lipid mediators generated from omega-3 fatty acids that are essential for tissue healing and the resolution of inflammation.

Ether Lipids: The structure of ether lipids is made up of an ether bond rather than an ester bond. They play a variety of roles in membrane dynamics and cellular signaling. Examples that stand out include:

- Platelet-activating Factor (PAF): PAF plays a role in bronchoconstriction, immunological responses, inflammation, and platelet aggregation.
- Ether-linked phospholipids: Found in cell membranes, these lipids function in cellular signaling cascades.

As more research is done, the classification of bioactive lipids continues to grow. The distinctive characteristics and capabilities of each class of bioactive lipids add to the complexity and variety of lipid-mediated cellular processes. Exploring bioactive lipids' functions in health, disease, and therapeutic interventions can be facilitated by having a solid understanding of how they are categorized (Gambino, 2018).

8.4 LIPID METABOLISM AND BIOSYNTHESIS

The production, disintegration, and alteration of lipids inside the body are part of the intricate processes known as lipid metabolism and biosynthesis. For cellular function, energy storage, membrane integrity, and the creation of bioactive lipid molecules to continue, these processes are essential. Understanding lipid metabolism and biosynthesis can help you better understand how lipid homeostasis is regulated and how it affects your general health. We explore the crucial facets of lipid metabolism and production in this section.

Lipogenesis

The process of creating fatty acids and triglycerides from non-lipid precursors is known as lipogenesis. Adipose tissue and other organs also participate in lipogenesis, although the liver is the main location. A number of enzymatic processes are involved, including as the production of fatty acids from acetyl-CoA, elongation of fatty acids, and esterification of fatty acids to form triglycerides. Several elements, including insulin, food intake, and hormonal cues, influence lipogenesis.

Lipolysis:

The conversion of stored triglycerides into glycerol and free fatty acids is known as lipolysis. This process, which occurs mostly in adipose tissue, is regulated by hormones like glucagon and adrenaline, which are located there In times of fasting or heightened energy needs, lipolysis offers a source of energy by releasing fatty acids that can be utilized for energy production by other tissues (Nicolaou, 2013).

Beta-oxidation:

The catabolism of fatty acids depends on the process of beta-oxidation. It involves the sequential conversion of fatty acids into acetyl-CoA units, which takes place in the mitochondria. In particular, during extended fasting or times of high energy demand, this mechanism produces ATP and serves as a significant source of energy for many tissues. The tight regulation of beta-oxidation is affected by nutritional and hormonal cues.

Cholesterol Biosynthesis:

Cholesterol is an essential lipid involved in a number of physiological activities, including the maintenance of cell membranes, the manufacture of hormones, and the creation of bile acids. Although other tissues also contribute, the liver is where cholesterol is largely made. The conversion of acetyl-CoA to mevalonate via the mevalonate route and subsequent enzymatic events leading to cholesterol production are the essential steps in cholesterol biosynthesis.

Phospholipid Biosynthesis:

Phospholipids play a key role in many cellular functions and are essential parts of cellular membranes. Fatty acids, glycerol, and other substances are put together to form phospholipid molecules during the process of phospholipid biosynthesis. Phosphatidic acid is produced during the synthesis of several phospholipids, including phosphatidylcholine, phosphatidylethanolamine, and phosphatidylinositol. The endoplasmic reticulum and other cellular compartments host enzyme activities that produce these phospholipids.

Sphingolipid Biosynthesis:

Sphingolipids are a varied class of lipids that are essential for membrane shape and cellular signaling. Ceramide is created when serine and palmitoyl-CoA condense to begin the process of sphingolipid biosynthesis. Ceramide is a precursor for the production of several sphingolipids, including glucosylceramide, gangliosides, and sphingomyelin. Multiple enzymatic processes take place in the endoplasmic reticulum and Golgi apparatus during the formation of sphingolipids (Nicolaou, A. 2013).

Eicosanoid Biosynthesis

Prostaglandins, thromboxanes, and leukotrienes are eicosanoids, which are bioactive lipid mediators made from polyunsaturated fatty acids like arachidonic acid. In order to create a variety of biologically active chemicals, cyclooxygenases (COX) or lipoxygenases (LOX) enzymatically oxidize fatty acids as part of the eicosanoid biosynthesis process. Inflammation, immunological responses, and a number of physiological functions all depend on these chemicals.

The optimal balance and usage of lipids inside the body are made possible by processes like lipid metabolism and production that are strictly regulated. Dysregulation of lipid metabolism can have a significant negative impact on health, increasing the risk of cardiovascular disease, metabolic illnesses, and other problems. New therapeutic targets and treatments for treating lipid-related illnesses are expected to be revealed by further investigation into the complex mechanisms of lipid metabolism and production (Gambino, 2018).

8.5 SIGNALING PATHWAYS AND RECEPTORS

Cells can receive and react to different stimuli, including bioactive lipids, through complex networks of molecular interactions called cellular signaling pathways. These signaling pathways are essential for information transmission and cellular response coordination, which help to maintain homeostasis and control physiological activities. Receptors are essential elements of these pathways because they identify particular signaling molecules and start subsequent signaling cascades. The principles of the signaling pathways and receptors involved in bioactive lipid signaling are examined in this section.

G Protein-Coupled Receptors (GPCRs):

The GPCR family of cell surface receptors is vast and is essential for bioactive lipid signaling. Numerous bioactive lipids bind to and activate particular GPCRs, including prostaglandins, lysophosphatidic acid, and sphingosine-1-phosphate (S1P), among others. GPCRs experience conformational changes upon ligand binding that result in the activation of intracellular signaling cascades. These pathways entail the activation of heterotrimeric G proteins, which alters the activity of downstream effectors such as adenylyl cyclase, phospholipase C, and ion channels, resulting in a variety of physiological reactions.

Nuclear receptors:

Nuclear receptors are transcription factors found in the nucleus that can bind to specific bioactive lipids like eicosanoids and some oxylipins. Bioactive lipids that activate nuclear receptors alter gene expression, which in turn affects a number of physiological processes. For instance, the nuclear receptors known as peroxisome proliferator-activated receptors (PPARs) control lipid metabolism, inflammation, and glucose homeostasis by binding to fatty acids and their derivatives.

Receptor tyrosine kinases (RTKs)

Some bioactive lipids, including as platelet-derived growth factors (PDGF) and specific eicosanoids, can activate the receptor tyrosine kinases (RTKs).

Transmembrane receptors with intrinsic kinase activity are known as RTKs. RTKs undergo autophosphorylation in response to ligand binding, which activates downstream signaling pathways like the Ras/MAPK pathway and the PI3K/Akt pathway. These routes control the proliferation, differentiation, growth, and survival of cells (Nicolaou, 2013).

Ion channels and transporters:

Ion channels and transporters can be modulated by certain bioactive lipids, such as endocannabinoids, which can impact cellular excitability and neurotransmission. Inhibition of neurotransmitter release can result from the activation of cannabinoid receptors on presynaptic terminals by endocannabinoids such as anandamide and 2-arachidonoylglycerol (2-AG).

Kinase cascades:

Bioactive lipids can also trigger kinase cascades, including the phosphoinositide 3-kinase (PI3K)/Akt pathway and the mitogen-activated protein kinase (MAPK) pathway. These cascades are involved in a number of biological functions, such as cell division, survival, and metabolism. For instance, the MAPK pathway can be activated by sphingosine-1-phosphate (S1P), which causes cellular migration and proliferation (Gambino, R. 2018).

Systems for second messengers:

Bioactive lipids have the ability to modify intracellular signaling pathways and function as second messengers. For instance, after the breakdown of phospholipids in the membrane, diacylglycerol (DAG) and inositol trisphosphate (IP3) are produced, which respectively cause the activation of protein kinase C (PKC) and the release of calcium ions. These second messengers take a role in the control of numerous physiological functions, such as neurotransmission and cell signaling.

Crosstalk and integration

Bioactive lipid-based signaling pathways frequently engage in crosstalk and integration with other signaling networks. These linkages enable intricate regulatory processes and customized cellular responses. There may be crosstalk between several bioactive lipid signaling pathways or between bioactive lipids and other signaling molecules, like cytokines or growth factors. Uncovering the physiological and pathological functions of bioactive lipid signaling requires an understanding of the signaling pathways and receptors involved. Dysregulation of these pathways may play a role in the emergence of a number of illnesses, such as cancer, inflammatory conditions, and metabolic disorders. The identification of new therapeutic targets for the treatment of lipid-related illnesses and the creation of more specialized and individualized therapies are predicted benefits of further investigation into these signaling systems.

8.6 ANALYTICAL TECHNIQUES IN BIOACTIVE LIPID RESEARCH

Understanding the roles, metabolism, and consequences of bioactive lipids in health and illness depends heavily on the analysis of these compounds, which is a fundamental component of lipid research. To recognize, measure, and describe bioactive lipids, a number of analytical methods have been developed. In this part, we examine a few of the frequently employed analytical methods in the study of bioactive lipids.

Mass Spectrometry:

Mass Spectrometry (MS) is a potent analytical method that is frequently used in lipid research. Based on their molecular weight and fragmentation patterns, it enables the identification and measurement of bioactive lipids. For different lipid classes and analytes, different MS techniques are used, such as liquid chromatography-mass spectrometry (LC-MS) and gas chromatography-mass spectrometry (GC-MS). MS-based methods make it possible to profile and quantify bioactive lipids in intricate biological samples, giving important information about lipid metabolism and signaling (Gambino, 2018).

Liquid chromatography (LC):

For the separation and analysis of bioactive lipids, liquid chromatography and mass spectrometry (LC-MS) are frequently used in conjunction. Depending on the lipid class and the desired qualities, several liquid chromatographic techniques, such as reverse-phase, normal-phase, and hydrophilic interaction chromatography, are used. High sensitivity and selectivity are made possible by LC-MS for the separation, identification, and quantification of bioactive lipids in complicated biological matrices.

Gas chromatography (GC)

Another popular method for examining bioactive lipids, particularly those with volatile or thermally stable characteristics, is gas chromatography. Based on their volatility and interactions with the stationary phase, GC separates lipid analytes. With GC, it is possible to examine oxylipins, fatty acids, and several lipid groups. Derivatization methods are frequently used to increase lipid volatility and detectability in GC analysis.

Thin-layer chromatography (TLC):

TLC is a simple and cost-effective technique for studying lipids. TLC differentiates lipid groups based on their various affinities for the stationary phase and migration on a thin layer of adsorbent material. It is commonly used to distinguish between different classes of lipids in samples and to identify lipids early. TLC can be used in combination with densitometry or other detection methods to quantify the separated lipids (Nicolaou, 2013).

NMR spectroscopy:

NMR spectroscopy is a powerful technique for analyzing the dynamics, composition, and structure of lipids. It provides information on lipids, intermolecular interactions, and the configuration of double bonds. Using NMR spectroscopy, it is possible to identify and measure bioactive lipids, investigate lipid-protein or lipid-membrane interactions, and more.

Imaging techniques:

The spatial detection and localization of bioactive lipids in tissues or cells is made possible by imaging techniques like fluorescence microscopy and matrix-assisted laser desorption/ionization imaging mass spectrometry (MALDI-IMS). These methods contribute to our understanding of the distribution, dynamics, and functions of lipids in biological systems, as well as how they impact cellular functions and the development of disease.

Enzyme-linked immunosorbent assay (ELISA):

The immunological method of enzyme-linked immunosorbent assay, or ELISA, is used to measure particular bioactive lipids or lipid mediators. It makes use of antibodies that firmly attach to the target lipid, enabling colorimetric or fluorescence detection for quantification. Prostaglandins, leukotrienes, and other lipid mediators implicated in inflammation and immunological responses are frequently tested by ELISA (Gambino, 2018).

Isotope tracing and metabolomics

Isotope tracing methods, such as radiolabeling and stable isotope labeling, in conjunction with mass spectrometry enable the analysis of lipid metabolism and turnover rates. The biosynthesis, breakdown, and interconversion of bioactive lipids in many biological systems can be better understood using isotope tracing. Lipidomics is one of the metabolomic methods that provide thorough lipid species profiling in biological samples, helping the identification of novel bioactive lipids and associated metabolic pathways.

Our knowledge of bioactive lipids and their functions in health and illness has substantially improved because to these analytical approaches. These methods enable researchers to decipher the intricate metabolism, signaling mechanisms, and activities of bioactive lipids, paving the door for the creation of novel treatment approaches and focused interventions in a range of lipid-related illnesses.

8.7 BIOACTIVE LIPIDS AND CELLULAR SIGNALING

Cells may receive extracellular signals and react to them thanks to a sophisticated network of molecular interactions called cellular signaling. These signals may come from the cell's immediate surroundings, hormones, growth factors, or other substances. These signals must be conveyed, which calls for precise coordination and internal communication. By interacting with particular receptors on the cell surface or inside the cell, bioactive lipids function as signaling molecules. These receptors, which include ion channels, nuclear receptors, receptor tyrosine kinases (RTKs), and G protein-coupled receptors (GPCRs), detect and react to the presence of bioactive lipids. These receptors undergo conformational changes after ligand binding, which activate the signaling pathways downstream (Nicolaou, 2013).

Bioactive lipids-based activation of signaling pathways results in a number of intracellular actions, including the activation or inhibition of enzymes, modifications to gene expression, and adjustments to ion fluxes. In the end, this results in the modification of cellular processes and physiological reactions. For instance, bioactive lipids from arachidonic acid, such as prostaglandins and leukotrienes, play important roles in immune responses and inflammation by activating certain receptors and triggering the generation of inflammatory mediators.

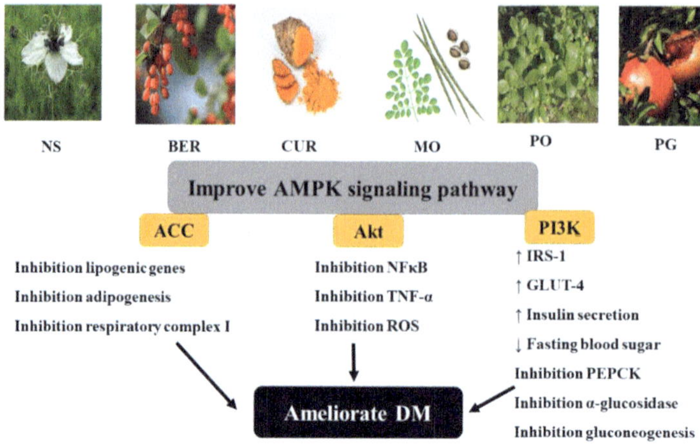

Figure 8.4. Effect of some natural plants on adenosine monophosphate-activated protein kinase (AMPK) signaling pathway. NS: *Nigella sativa*, BER: berberine, CUR: curcumin, MO: *Moringa olifera*, PO: *Portulaca oleracea*; PG: *Punica granatum*, ACC: acetyl CoA carboxylase, Akt: protein kinase B, NF-κB: nuclear factor-kappa B, TNF-α: tumor necrosis factor-alpha, ROS: reactive oxygen species, PI3K: phosphatidyl inositol-3-kinase, IRS-1: insulin receptor substrate-1, GLUT-4: glucose transporter-4, FBS: fasting blood sugar, PEPCK: phosphoenolpyruvate carboxykinase, α-glucosidase: alpha-glucosidase, DM: diabetes mellitus.

Source: Balbaa M, El-Zeftawy M, Abdulmalek SA. Therapeutic Screening of Herbal Remedies for the Management of Diabetes. *Molecules*. 2021; 26(22):6836. https://doi.org/10.3390/molecules26226836 [accessed 6 Jun, 2023]

Additionally, bioactive lipids control how well cells grow, divide, and survive. For example, phosphatidylinositol phosphates trigger the phosphatidylinositol 3-kinase (PI3K)/Akt pathway, which is essential for cell growth and survival. Through the activation of GPCRs and RTKs, sphingolipids like sphingosine-1-phosphate (S1P) are associated with cell migration, angiogenesis, and tissue remodeling. Dysregulation of bioactive lipid signaling pathways may be the initial cause of many diseases, including cancer, cardiovascular problems, inflammatory diseases, and metabolic disorders. Understanding the intricate interaction between bioactive lipids and cellular signaling is essential to pinpoint the root causes of disease and discovering new treatment targets. Through their interactions with certain receptors and activation of intracellular signaling pathways, bioactive lipids function as important signaling molecules that modify cellular responses.

Their participation in a variety of physiological processes highlights their significance in preserving cellular homeostasis and controlling many facets of cell function. The pursuit of new knowledge in this area promises to open the door to the creation of novel therapeutic strategies and shed light on the functions of bioactive lipids in cellular signaling (Gambino, 2018).

Sphingolipids: Sphingolipids are a group of bioactive lipids that are crucial for many physiological functions, such as cell signaling, membrane structure, and cell-cell communication. A sphingoid base, a fatty acid, and a polar head group make up these complex lipids. They can be found in all eukaryotic cells, but the nervous system has the highest concentration. Sphingolipids play a variety of roles in cellular processes because of their structural variety and capacity to operate as signaling molecules. They can be found in the plasma membrane, endoplasmic reticulum, Golgi apparatus, and lysosomes, among other cellular compartments. We go through some of the main features of sphingolipids and their biological importance here:

Structure and Classification: Polar head groups on sphingolipids are used to categorize them. Ceramides, sphingomyelins, glycosphingolipids (including cerebrosides and gangliosides), and sphingosine-1-phosphate (S1P) are a few examples of typical sphingolipids. The variety of sphingolipid types adds to the complexity of cellular operations because each type has a distinct structure and purpose. Sphingolipids, which are important parts of cell membranes, help to organize and maintain the fluidity of the membrane. Sphingolipids have unique features that affect the fluidity, stability, and curvature of cell membranes, such as lengthy hydrocarbon chains and the presence of the sphingoid backbone. The functioning of membrane proteins as well as cellular procedures including membrane trafficking and signal transduction are thus impacted (Nicolaou, 2013).

Sphingolipids serve as significant signaling molecules in several biological processes. Ceramides, for instance, control the cell cycle, cell differentiation, and death. They can activate particular protein kinases and phosphatases, which modulate biological reactions. Sphingosine-1-phosphate (S1P) controls a variety of processes, including cell migration, angiogenesis, immunological responses, and vascular tone, by acting as a bioactive lipid mediator and activating particular G protein-coupled receptors.

Sphingolipids are important players in the signaling and stress response processes that occur within cells. Ceramides are produced in response to cellular stresses like oxidative stress or exposure to certain toxins, and they function as signaling molecules to control cell fate decisions including apoptosis and autophagy. Ceramides and other sphingolipids are in a delicately regulated equilibrium that can affect how cells react to stress and disease processes.

Dysregulation of sphingolipid metabolism and signaling has been connected to a number of illnesses. For instance, altered sphingolipid metabolism is linked to neurodegenerative diseases including Parkinson's and Alzheimer's disease. Inherited diseases including Gaucher disease and Niemann-Pick disease can result from mutations in enzymes involved in sphingolipid metabolism. Furthermore, abnormal sphingolipid signaling has been connected to diabetes, cardiovascular disease, cancer, and inflammation.

Sphingolipids play a role in many disorders, therefore targeting their metabolism and signaling pathways has been identified as a viable treatment strategy. For the development of innovative therapeutics, targeting receptors engaged in sphingolipid signaling pathways or modifying the levels or activities of particular enzymes involved in sphingolipid metabolism show promise. Sphingolipids are a varied group of bioactive lipids that are essential for many physiological functions, such as membrane shape, cell signaling, and stress response. The significance of comprehending their roles and control is highlighted by their involvement in a variety of physiological and pathological processes. The development of innovative therapeutic approaches for a variety of disorders is predicted to benefit greatly from more investigation into sphingolipid metabolism and signaling pathways.

Eicosanoids: Eicosanoids are a group of bioactive lipids generated from the polyunsaturated fatty acid arachidonic acid. These lipid mediators are essential for controlling physiological processes such as vascular function, immunological responses, inflammation, and others. Eicosanoids act locally and are synthesized on demand in response to a variety of stimuli. They can have either paracrine or autocrine actions. They assist in preserving homeostasis and directing cellular reactions in a variety of body tissues.

Prostaglandins, leukotrienes, and thromboxanes are the three main families of eicosanoids based on their chemical makeup and biological roles. Prostaglandins (PGs), which come in a variety of subtypes like PGE2, PGD2, and PGF2, are produced by the cyclooxygenase (COX) pathway. Leukotrienes (LTs), which include LTB4 and the cysteinyl leukotrienes (LTC4, LTD4, and LTE4), are produced by the 5-lipoxygenase (5-LOX) pathway. TXA2 is one of the thromboxanes (TXs) produced by the COX pathway (Nicolaou, 2013).

Eicosanoids are well-known for playing important roles in inflammation. Immune cells that are stimulated during inflammation release arachidonic acid from cell membrane phospholipids, including mast cells and macrophages. The COX and LOX enzymes then break down arachidonic acid to produce prostaglandins and leukotrienes, respectively. Vasodilation, enhanced vascular permeability, and the attraction of immune cells to the site of inflammation are all influenced by prostaglandins like PGE2. Leukocyte migration and activation are encouraged by leukotrienes, especially LTB4, which intensifies the inflammatory response.

Eicosanoids play critical functions in platelet aggregation and vascular and platelet function. By encouraging platelet aggregation and vasoconstriction, thromboxane A2 (TXA2), which is produced by platelets, plays a role in hemostasis and thrombosis. In contrast, the endothelial cell-produced prostacyclin (PGI2) has the opposite effects, acting as a powerful vasodilator and a platelet aggregation inhibitor. For the maintenance of vascular homeostasis, the interaction between TXA2 and PGI2 must be in balance (Gambino, 2018).

Eicosanoids are involved in a variety of reproductive processes, such as menstruation, ovulation, and parturition. During labor and delivery, prostaglandins, in particular PGE2 and PGF2, are implicated in uterine contractions and cervical dilatation. Clinically, prostaglandin production inhibition is frequently used to prevent premature labor.

LTC4, LTD4, and LTE4 are leukotrienes that have a substantial impact on the respiratory system. They are strong bronchoconstrictors and have a significant impact on the development of allergic respiratory disorders like asthma. A typical therapeutic approach for treating asthma is to inhibit leukotriene production or block its receptors.

Targeting eicosanoid pathways has therapeutic ramifications because they participate in numerous physiological and pathological processes. Nonsteroidal anti-inflammatory medicines (NSAIDs) reduce pain and inflammation by blocking the COX enzymes, which in turn lowers the production of prostaglandins. To treat inflammatory and pulmonary diseases, medications that target leukotriene receptors specifically or obstruct leukotriene synthesis have been created (Nicolaou, A. 2013).

The wide class of bioactive lipids known as eicosanoids, which are generated from arachidonic acid, are essential for inflammatory reactions, immunological responses, vascular function, and other physiological activities. Their role in preserving homeostasis and directing cellular responses is highlighted by their capacity to serve as local mediators in different tissues. The development of tailored therapeutics can benefit from an understanding of the biology of eicosanoids and how their dysregulation in disease states.

Endocannabinoids: The body's ligands for cannabinoid receptors are a class of bioactive lipids known as endocannabinoids. They are produced by the body naturally and are essential for many physiological processes, including the regulation of pain, appetite, immunity, mood, and neuroprotection. Endocannabinoids and their receptors make up the endocannabinoid system, which controls numerous biological processes and protects homeostasis.

Endocannabinoids come in many different forms, but anandamide (AEA) and 2-arachidonoylglycerol (2-AG) are the two that have received the most research. While 2-AG is created from the precursor molecule diacylglycerol (DAG), anandamide is produced from the fatty acid arachidonic acid.

Endocannabinoids work by attaching to and activating cannabinoid receptors, mainly cannabinoid receptor types 1 (CB1) and 2 (CB2). While CB2 receptors are mainly present in immune cells and peripheral organs, CB1 receptors are mainly prevalent in the central nervous system, including the brain. These G protein-coupled receptors are a group of receptors that, when activated, modify many signaling pathways.

Endocannabinoids and their receptors are required for a variety of physiological processes. For instance, they assist in regulating the nervous system's synaptic plasticity, neuronal excitability, and neurotransmitter release. Endocannabinoids have a role as local mediators to block pain signals and reduce inflammation. They are also engaged in the control of inflammatory and pain responses. Endocannabinoids affect digestive tract metabolism, appetite, and food intake. The endocannabinoid system has also been connected to the regulation of mood, memory, and stress (Gambino, 2018).

Regulation and breakdown: "On-demand" production and breakdown of endocannabinoids is a procedure that carefully controls the activity of these compounds. After being released, they are quickly absorbed by cells and degraded by certain enzymes. While 2-AG is predominantly metabolized by the enzyme monoacylglycerol lipase (MAGL), anandamide is primarily metabolized by the enzyme fatty acid amide hydrolase (FAAH). Endocannabinoid signaling is preserved as a temporary and local effect as a result.

Therapeutic Potential: Due to its therapeutic potential for treating a variety of illnesses, the pharmacological manipulation of the endocannabinoid system has attracted a lot of attention. The use of exogenous cannabinoids (like THC and CBD), synthetic cannabinoid analogs, or modulators of endocannabinoid metabolism as part of cannabinoid-based therapies has been investigated for the treatment of chronic pain, neurodegenerative diseases, epilepsy, mood disorders, and conditions related to inflammation. In addition, the endocannabinoid system is being researched as a potential pharmacological target for conditions like cancer, metabolic disease, and immune-related illnesses (Nicolaou, 2013).

Clinical Relevance: Many clinical disorders have been linked to disruption of the endocannabinoid system. Mood disorders, neurodegenerative diseases like Alzheimer's and Parkinson's disease, and chronic pain syndromes, for instance, have been linked to changes in endocannabinoid signaling. The creation of new treatment approaches may result from a better understanding of the complex interactions between endocannabinoids and their receptors in various disease conditions.

Numerous physiological functions in the body are controlled by endocannabinoids and the endocannabinoid system. They interact with cannabinoid receptors to affect several physiological and pathological processes, including pain, hunger, immunological response, mood, and neuroprotection. New therapeutic strategies aimed at this complex signaling system could be developed with more study of the endocannabinoid system.

Lysophospholipids: A group of bioactive lipids known as lysophospholipids are crucial for cellular signaling and physiological functions. The phospholipase enzymes, which operate on phospholipids to remove one fatty acid chain from the glycerol backbone and produce a single acyl chain and a polar head group, are responsible for their production. Lysophospholipids are widely distributed in cell membranes and are essential for some cellular processes.

Lysophospholipid kinds: Lysophosphatidic acid (LPA) and lysophosphatidylcholine (LPC) are the two most prevalent lysophospholipid kinds. While LPC is made from phosphatidylcholine, LPA is produced from phosphatidic acid. These lysophospholipids contribute to a variety of cellular activities with their unique structures and roles.

Lysophospholipids can bind to particular G protein-coupled receptors (GPCRs) on the cell surface and function as signaling molecules. For instance, LPA interacts with LPA receptors and initiates intracellular signaling cascades that control cell motility, proliferation, and survival as well as cytoskeletal reorganization. These receptors' activation can influence many physiological processes, including cancer development, inflammation, and wound healing (Gambino, 2018).

Lysophospholipids, notably LPC, have been linked to inflammatory reactions as mediators of inflammation. LPC may stimulate immune cells like neutrophils and macrophages, which causes them to produce cytokines and inflammatory mediators. Additionally, lysophospholipids can influence vascular permeability and encourage the migration of immune cells to areas of inflammation, both of which contribute to the inflammatory process.

Lysophospholipids and cardiovascular health and disease have been connected. LPA, for instance, may encourage smooth muscle cell proliferation and vasoconstriction, which may help to develop cardiovascular diseases. Blood clotting and the health of the vascular system may be affected by lysophospholipids' effects on platelet function and thrombosis. Lysophospholipids, particularly LPA, have been connected to many elements of the onset and development of cancer. LPA can encourage the growth, migration, invasion, and resistance to apoptosis of cancer cells. Additionally, it can affect angiogenesis, promoting the establishment of new blood vessels to support tumor development. The aggressiveness and spread of various cancer types have been linked to the deregulation of lysophospholipid signaling pathways (Nicolaou, 2013).

Due to the numerous physiological and pathological processes that lysophospholipids are engaged in, addressing their signaling pathways has emerged as a promising therapeutic approach. Lysophospholipid receptor modulators and enzymes involved in lysophospholipid metabolism are potential therapeutic targets for diseases like cancer, cardiovascular disease, and inflammatory disorders. Lysophospholipids, a class of bioactive lipids, are essential for cellular signaling and physiological processes. Their part in inflammation, cardiovascular health, cancer, and other diseases serves as a reminder of their significance in human biology. Future research suggests that lysophospholipids may be used in novel therapeutic methods to address a range of diseases.

8.8 BIOACTIVE LIPIDS AND INFLAMMATION

The body's reaction to injury, infection, or tissue damage is inflammation, which is greatly influenced by bioactive lipids. Many immune cells, cytokines, and molecular mediators are all involved in the complicated process of inflammation. Bioactive lipids function as strong signaling molecules during inflammation, controlling inflammatory mediator synthesis, vascular permeability, and immune cell activation.

Prostaglandins:

The cyclooxygenase (COX) enzymes transform arachidonic acid into prostaglandins. They play a critical role in inflammation by regulating vasodilation, vascular permeability, and the recruitment of immune cells to the site of inflammation. Redness, swelling, and discomfort associated with inflammation are caused by prostaglandins such as PGE2. They also encourage the synthesis of the inflammatory mediators' chemokines and cytokines.

Leukotrienes:

Through the action of the enzyme 5-lipoxygenase (5-LOX), leukotrienes are produced from arachidonic acid. They are effective inflammatory mediators, especially in allergy and asthmatic reactions. Leukotrienes, including LTB4, encourage the attraction and activation of immune cells, including neutrophils and eosinophils, which intensifies the inflammatory response. In the respiratory system, other leukotrienes such as LTC4, LTD4, and LTE4 cause bronchoconstriction, increased vascular permeability, and mucus formation (Gambino, 2018).

Platelet-Activating Factor (PAF):

A bioactive lipid generated from phospholipids called platelet-activating factor (PAF) is important in inflammatory reactions. Numerous immune cells, such as neutrophils, macrophages, and mast cells, generate it. By encouraging platelet aggregation, vasodilation, and enhanced vascular permeability, PAF functions as a powerful inflammatory mediator. Additionally, it promotes cytokine and chemokine production, which exacerbates inflammation (Nicolaou, 2013).

Sphingolipids:

Sphingolipids, such as sphingosine-1-phosphate (S1P), have come to be recognized as significant inflammatory regulators. During inflammation, S1P functions as a signaling protein that controls immune cell movement and activity. It can facilitate immunological responses by promoting immune cell egress from lymphoid organs into the circulation and tissues. Through intracellular signaling pathways, sphingolipids also control the synthesis of inflammatory mediators like cytokines and prostaglandins.

Endocannabinoids:

It has been demonstrated that endocannabinoids can control inflammation, including anandamide and 2-arachidonoylglycerol (2-AG). They produce anti-inflammatory effects by interacting with immune cells' cannabinoid receptors, especially CB2 receptors. Endocannabinoids can prevent the migration of immune cells, reduce the generation of pro-inflammatory cytokines, and aid in the healing of inflammation.

Resolvins and Lipoxins:

Specialized pro-resolving lipid mediators called lipoxins and resolvins are essential for controlling inflammation. They come from omega-3 fatty acids and actively encourage tissue healing, the reduction of inflammation, and the reestablishment of homeostasis. Lipoxins and resolvins block the effects of pro-inflammatory mediators, lower the recruitment of immune cells, and improve the removal of inflammatory byproducts.

Having a better understanding of how bioactive lipids influence inflammation can aid in the development of therapeutic strategies that focus on these molecules. . As prospective anti-inflammatory treatments, researchers are looking into receptor antagonists, intracellular signaling pathway modulators, and enzyme inhibitors of bioactive lipid synthesis. Utilizing the complex signaling pathways of bioactive lipids may result in the creation of fresh therapies for inflammatory conditions such as inflammatory bowel disease, arthritis, asthma, and cardiovascular problems (Nicolaou, 2013).

8.9 BIOACTIVE LIPIDS AND METABOLIC DISORDERS

In addition to being connected to the initiation and development of metabolic disorders, bioactive lipids are increasingly recognized as essential metabolic process regulators. These lipids play a crucial role in regulating insulin sensitivity, lipid metabolism, energy balance, and inflammation, all of which are crucial for preserving metabolic health (Gambino, 2018). Insulin resistance and obesity have a substantial impact on metabolic diseases type 2 diabetes and cardiovascular disease. Long-chain fatty acids, ceramides, and diacylglycerols (DAGs) are a few bioactive lipids that have been connected to the pathophysiology of insulin resistance brought on by obesity. In organs like the liver, skeletal muscle, and adipose tissue, these lipids may accumulate and impede insulin signaling and glucose uptake. Elevated levels of ceramide and DAG cause disruption of the insulin signaling pathways, leading to insulin resistance and delayed glucose metabolism.

Lipid Dysregulation and Dyslipidemia: Metabolic illnesses like obesity, type 2 diabetes, and metabolic syndrome are frequently accompanied by dyslipidemia, which is defined by high levels of triglycerides and low-density lipoprotein cholesterol. Lysophosphatidylcholine (LPC) and oxidized lipids are two bioactive lipids that have been linked to the emergence of dyslipidemia. A good example of how LPC might contribute to dyslipidemia. is by encouraging the production of pro-inflammatory cytokines, which increases lipogenesis and decreases lipolysis.

Metabolic diseases are characterized by persistent low-grade inflammation, and bioactive lipids are crucial in controlling inflammatory responses. Prostaglandins and leukotrienes, two lipid mediators generated from arachidonic acid, can encourage inflammation and insulin resistance. Furthermore, adipokines like adiponectin and leptin, which are produced from adipose tissue, control the activity of immune cells and inflammatory reactions. Systemic inflammation and metabolic dysfunction may be exacerbated by adipokine dysregulation.

The regulation of metabolism has been attributed to the endocannabinoid system, which consists of endogenous cannabinoids (endocannabinoids) and their receptors. Anandamide and 2-arachidonoylglycerol (2-AG) are examples of endocannabinoids that can affect insulin sensitivity, lipid metabolism, energy expenditure, and food intake. Obesity, insulin resistance, and dyslipidemia have been linked to endocannabinoid system dysregulation. It has become clear that one potential therapeutic strategy for metabolic disorders is to target this mechanism.

Bioactive lipids are important regulators of metabolism and contributors to the emergence of metabolic diseases. Insulin resistance, dyslipidemia, inflammation, and other metabolic disorders are all influenced by the dysregulation of these lipids. It becomes possible to create novel treatment strategies to treat metabolic illnesses and enhance general metabolic health by comprehending the intricate interactions between bioactive lipids and metabolic pathways (Gambino, 2018).

Chapter

9 POTENTIAL APPLICATIONS OF BIOACTIVE MOLECULES

CONTENTS

Plants have several physiologically active non-nutrients. Bioactive substances produce a biological reaction in humans and animals. Bioactive foods are useful foods. Functional foods are those with health advantages beyond basic nourishment for the intended demographic. To avoid chronic illnesses, new food items with higher bioactive component content are created. Optimized methodologies and methods for bioactive chemical separation, detection, valuation, and biological impacts are needed to comprehend their advantages. To prove health benefits from dietary components, bioactive chemical identification and quantification must be accurate and reproducible. This is crucial to determine suitable food consumption levels and safety limits for validating health claims and obtaining targeted health benefits (Zuorro, 2021).

Bioactive compounds are abundant in fruits, vegetables, and byproducts obtained from them. These bioactive compounds are used in a variety of sectors, including the pharmaceutical, dairy, and chemical industries. These bioactive must first be isolated from their natural matrices, evaluated, and classified before they may be used in particular applications. They can only be used when these steps have been completed. Because active components sourced from natural sources are required, extraction processes that are both acceptable and standardized are required. Subcritical water extraction (SWE), enzyme-aided extraction (EAE), microwave-assisted extraction (MAE), and ultrasound-assisted extraction (UAE) are new methods for extracting bio actives that are acceptable, safe, cost-effective, time-consuming, and environmentally friendly. Because of their low energy consumption, reduced effluent output, reduced water consumption, reduced reliance on potentially toxic solvents, and ecological soundness, these techniques are ecologically benign and sustainable. Traditional methodologies and cutting-edge technology, as well as their underlying theoretical notions, values, and limitations, are contrasted in this examination to obtain the target content (Fernandes da Silva, 2017)

Natural materials like plant extracts offer countless opportunities for medication discovery because of their chemical variability. According to the WHO, traditional medicine serves as the primary form of care for 80% of people worldwide. Asian herbal medicine is a long-standing human-environment interaction. Traditional medicine plants include several compounds that cure chronic and infectious ailments. Men turned to ethnopharmacology once chemically created medications caused side effects and microbial resistance. They uncovered hundreds of safe, useful phytochemicals from plants. Anticancer, antibacterial, antioxidant, antidiarrheal, analgesic, and wound-healing activities were reported. People often say natural or herbal products are beneficial. To prove this conventional claim, bioactive chemical clinical studies are needed. New bioactive chemicals and their formulations (extracts) must be carefully evaluated for pharmacokinetics, bioavailability, effectiveness, safety, and medication interactions in clinical studies. Before the treatment is widely used, clinical studies are carefully structured to protect participants and solve particular research objectives by testing for immediate and long-term negative effects.

Plants have traditionally provided food and medicine. Herbivores and omnivores evolved because they provide nutrition and medications. Secondary metabolites have specific functions. These chemicals protect plants from abiotic and biotic stress. Plant's chemical molecular structures are as complicated as their variety and chemical diversity, and their biosynthetic and metabolic processes are harmoniously intertwined, demonstrating their adaptability to changing environmental conditions. Various plant meals have distinct bioactive chemicals and effects. Plant-based diets provide a wide spectrum of bioactive compounds and health advantages (Fernandes da Silva,2017).

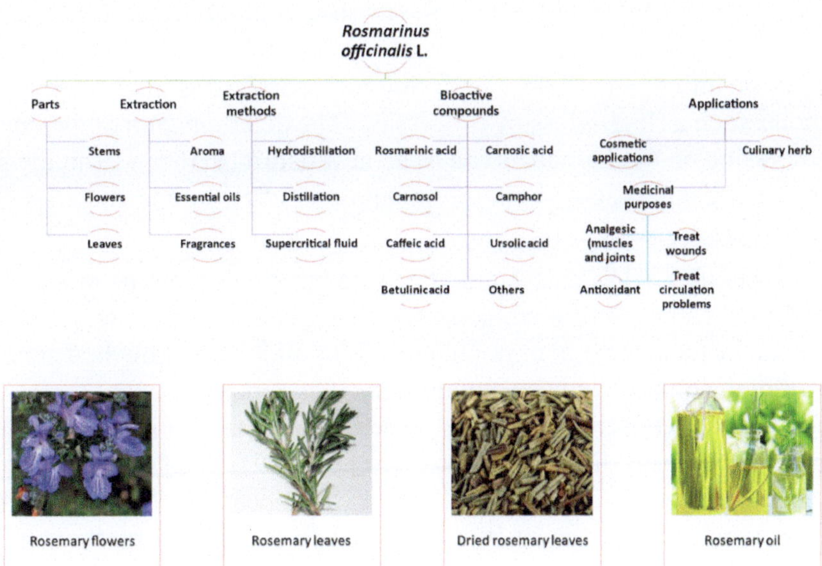

Figure 9.1. Bioactive compounds from rosemary.

Source: El Boukhari R, Fatimi A. A Review of the Patentability of Rosemary-Derived Drugs and Bioactive Compounds. *Drugs and Drug Candidates*. 2023; 2(1):172-188. https://doi.org/10.3390/ddc2010011 [accessed 5 Jun, 2023]

9.1 ANTIOXIDANT ACTIVITY

The examination of the levels of natural antioxidants present in plant-based foods is a topic of significance in the context of human health, as these compounds have the potential to contribute to the promotion of overall well-being. Additionally, the investigation of endogenous antioxidant components in terms of their ability to prolong the shelf life of food products is also of interest. This article provides a comprehensive discussion of the mechanism of antioxidant action and the various techniques utilized to measure its activity. It also delves into the different classes of phenolic compounds, namely phenolic acids, flavonoids, lignans, stilbenes, and tannins, as well as the plant sources of antioxidants, which include oil seeds, cereals, legumes, plants in the Lamiaceae, tea and coffee, tree nuts, fruits, and berries. Additionally, the article explores the various extraction techniques employed to obtain phenolic compounds from plant material, and the effects of processing and storage on the natural antioxidant content of food. Under optimal circumstances, thermal processing results in a decrease in antioxidant capacity or exhibits negligible adverse impacts. This phenomenon holds even in instances where phenolic compounds that are bound remain unextracted from the food's structural matrix. The antioxidant capacity of food is more significantly reduced by thermal procedures, such as steaming, simmering, or frying, compared to gentler sterilization methods, such as high-pressure processing. The determination of the impact of processing on food products can be achieved through the measurement of their antioxidant capacity during formulation or after storage under specific conditions (Zuorro, 2021).

The cellular redox state is modified through direct neutralization of free radicals and chelation of metal ions, which are known to enhance oxidative reactions. Pesticides are commonly utilized to inhibit the growth of fungi in agricultural, culinary, and feed commodities. The employment of fungicides and synthetic preservatives has raised concerns among consumers, researchers, and stakeholders, owing to the potential risks of chronic toxicity, carcinogenicity, and environmental harm. The appearance of fungal strains that are resistant to treatment is a significant factor that warrants attention. As a result, it's feasible that stakeholders and customers will be more receptive to using natural resources. Plant metabolism generates bioactive compounds that exhibit a multifaceted chemical profile and perform diverse biochemical and physiological roles. As a result, these molecules are deemed to be adaptive. The precise and unambiguous function of plant secondary metabolites presents a challenge (Fernandes da Silva, 2017).

Plants are considered to be the most optimal source of bioactive compounds. Bioactive metabolites are classified into primary or secondary categories based on their respective functions. Plants and fungi synthesize diverse secondary metabolites in response to various factors such as physiological stage, tissue location (including floral and non-floral leaves, fruits, and bark), environmental conditions, and biotic or abiotic stress. The aforementioned compounds are believed to potentially exert an influence on the fundamental physiological processes of the cell. In their capacity as phytohormones, they could potentially participate in the regulation of cellular development and growth. Several crucial plant hormones have been identified, including ethylene, auxin, gibberellins, abscisic acid, cytokinins, brassinosteroids, and polyamines. Their main purpose is ecological, particularly in terms of safeguarding plants against herbivores, bacteria, and fungi (Zuorro, 2021).

Plants utilize diverse defense mechanisms to resist the invasion of diseases. The aforementioned systems are dependent on either anatomical or biochemical characteristics, which are categorized as passive and constitutive defense, or on active pathogen-induced modifications, which fall under the active and inducible defense categories. Terpenes are capable of being synthesized at basal levels and can be induced to express higher levels in response to tissue damage or pathogen infection. Phytoalexins, pathogen-related (PR) proteins, chitinases, and glucanases are among the constituents of inducible compounds. Passive or constitutive defense chemicals, such as glucosides, saponins, antifungal proteins, enzyme inhibitors, and antifeedants, are present in various organisms as a means of defense. The plant is capable of synthesizing and storing glycosylated or conjugated defense-related metabolites in a non-toxic state, which allows for their presence. The implementation of conjugation or precise localization within vacuoles or other subcellular compartments are techniques utilized to prevent auto allelopathy and facilitate the prompt production of active forms.

In contrast, certain plant species produce antifungal substances during an infection. The molecules known as phytoalexins exhibit similarities to constitutive antifungal toxins, albeit with a higher degree of lipophilicity. Phytoecdysones, which are endogenous compounds synthesized by plants and exhibit endocrine activities in animals, have the potential to modulate or expedite insect ontogeny. They can form symbiotic associations with fungi and lichens. There exists a plethora of techniques for the extraction of bioactive compounds, which include: Soxhlet extraction, maceration, and hydro distillation are established conventional methods. Various methods such as ultrasounds, microwaves, electric fields, high pressures, or supercritical fluids have been explored to reduce solvent usage and establish ecologically sustainable extraction conditions, as reported in studies. The process of eliminating residual water from a substance can result in the production of a highly concentrated substance referred to as an essential oil (EO). Secondary metabolites are classified into three primary categories based on their biosynthetic origin. The three categories of compounds under consideration are phenolics, terpenes, and nitrogen-containing compounds (Fernandes da Silva, 2017).

9.2 ANTI-INFLAMMATORY EFFECTS

Inflammation, a pathological condition, is associated with numerous diseases, including malignancy, cardiovascular disease, and chronic inflammatory disorders. In addition, the concentrations of inflammatory biomolecules in the circulation are higher in obese individuals with robust physical constitutions. In addition to nutrients, food contains bioactive substances that regulate inflammatory and immunomodulatory processes but are not nutrients. According to empirical data, nutrition has a substantial effect on the body's inflammatory systems. The consumption of fruits, vegetables, and whole grains has been found to significantly reduce the likelihood of inflammation. Some plant-based diets may have anti-inflammatory properties, according to human intervention studies and observational research. Carotenoids and flavonoids, bioactive components of plant-based diets, have been found to influence the inflammatory and immune systems. In conclusion, there is substantial evidence that diets rich in plant-based foods and the non-nutritive components of such meals influence inflammatory and immunological responses. A diet rich in plant-based foods has the potential to reduce the risk of developing cancer and cardiovascular disease via the mechanism of reduced inflammation. Fruits, vegetables, and whole cereals are recommended by all major international nutrition organizations due to their rich diversity of bioactive compounds and health-promoting properties.

Chronic low-grade inflammation is associated with chronic non-communicable diseases (CNCDs). Bioactive substances can alleviate mild inflammation. This review focuses on the relationship between noncommunicable diseases, inflammation, and the ingestion of catechins, curcumin, quercetin, and resveratrol (Zuorro, 2021). Due to their potential use in disease prevention and treatment, plants with therapeutic properties are gaining importance in the agricultural and pharmaceutical industries. In one study, colorimetric, chromatographic, and spectrophotometric techniques are used to determine the phenolic profile and antioxidative potential of seven medicinally and nutritionally significant plants. Among the botanicals being examined are Salvia *officinalis L., Rosmarinus officinalis L., Olea europaea L., Punica granatum L.*, and the leaves and juvenile stems of *Ruta graveolens L.* The highest concentrations of ortho-diphenols (391.76 mg gallic acid/g of plant dry weight), tannins (99.20 mg epicatechin/g of plant dry weight), and flavonoids (24 mg catechin/g of plant dry weight) were found in the hydro-methanolic leaf extracts of *P. granatum*, also known as the pomegranate. The ABTS, DPPH, and FRAP experiments revealed that the antioxidant capacities of pomegranate leaf extracts were significantly greater than those of other species (2.14, 2.27, and 2.33 mM Trolox per gram of plant dried weight, respectively). The observed values were found to be 4-200 times greater than those of other species. The antioxidant properties of pomegranate leaves are due to their diverse phenolic components and high tannin content, whereas the primary phenolic classes in the other six plants were discovered to be phenolic acids and flavonoids. Pomegranate leaves may be a valuable botanical resource for developing novel functional food-pharmaceutical products that are abundant in natural bioactive constituents. In addition to possibly improving human health, this may also contribute to the biodiversity of the ecosystem. Numerous plant species possessing antioxidant properties is a well-known fact (Fernandes da Silva, 2017).

9.3 CARDIOVASCULAR HEALTH

Because of their high nutritious content and calorie density, plant-based meals are extensively consumed across the world. The rise in metabolic illnesses associated with animal product consumption has increased their consumption. In addition to their nutritional benefits, plant-based foods include a varied spectrum of bioactive chemicals that promote health. Plant-derived bioactive substances, such as proteins, polyphenols, phytosterols, biogenic amines, carotenoids, and others, have been demonstrated in studies to benefit human health. These advantages include cancer prevention and treatment, cardiovascular disease prevention and treatment, and diabetes prevention and treatment, as well as improvements in gastrointestinal, immunological function, and neurological problems. Previous research has shown that bioactive chemicals have antioxidative, anti-inflammatory, and immunomodulatory properties, as well as the ability to improve intestinal barrier integrity. These features, when combined, allow bioactive compounds to reduce the negative consequences of a variety of human illnesses. The chapter examines the processes behind the bioactive qualities of various plant components, including bioactive components derived from fruits, vegetables, grains, and other plant sources with health-promoting effects. The primary focus of such research is on the examination of bioactive substances derived from food, intending to offer significant insights to researchers in the development of novel ways. This includes identifying and selecting promising bioactive elements for the production of functional meals that may be used in the management of a variety of noncommunicable disorders (Zuorro, 2021).

Cardiovascular diseases (CVDs) are responsible for 30% of all fatalities globally, making them the leading cause of death. According to World Health Organization (WHO) research, the yearly death rate from cardiovascular diseases (CVDs) is predicted to be 17.9 million, with a projected increase to 22.2 million by 2030. Individuals' mortality rates rise as they age. Females are more likely than men to die from cardiovascular disease, with a frequency of 51% and 42%, respectively. Many people use traditional plant medicine, also known as phytochemicals, as an alternative or supplement to commercially available medications to alleviate and prevent cardiovascular disease (CVD). The current study intends to investigate the efficacy of 92 plant species, 15 of which are terrestrial plants, in the treatment of cardiovascular disease (CVD). Commonly used botanicals in the treatment of cardiovascular disease include *Daucus carota*, *Nerium oleander*, *Amaranthus viridis*, *Ginkgo biloba*, *Terminalia arjuna*, *Picrorhiza kurroa*, *Salvia miltiorrhiza*, *Tinospora cordifolia*, *Mucuna pruriens*, *Hydrocotyle asiatica*, *Bombax ceiba*, and *Andrographis paniculata*. Flavonoids, polyphenols, plant sterols, plant sulfur compounds, terpenoids, and other bioactive phytochemicals are found in the aforementioned flora. The reduction of low-density lipoprotein oxidation, which increases vasodilation, is a well-reported mode of action of flavonoids.

Plant sterols have been discovered to be protective against cardiovascular disease due to their ability to prevent blood cholesterol absorption. Plant sulfur compounds have been demonstrated to reduce the incidence of CVD by activating nuclear factor-erythroid factor 2-related factor 2 (Nrf2) and blocking cholesterol production. Quinone helps to reduce the risk of cardiovascular disease by increasing ATP production in mitochondria. Terpenoids, on the other hand, work to slow the progression of atherosclerotic lesions in the aortic valve. Numerous plant species have been discovered to have a wide range of physiologically active ingredients with well-established biological effects. Nonetheless, there is still a need for cardiovascular disease prevention and treatment. More research is needed to acquire a complete knowledge of the mechanism and phytochemical contents of certain botanicals used in the treatment of cardiovascular diseases (Fernandes da Silva, 2017).

Medicinal botanicals have been discovered to exert cardioprotective effects on a variety of cell types, including macrophages and monocytes, cardiomyocytes, vascular smooth muscle cells, and endothelial cells. The cardiovascular system has been shown to benefit from anti-inflammatory, anticoagulant, hypolipidemic, hypotensive, and diuretic effects. The principal method of reducing the consequences of cardiovascular disease is through antioxidant mechanisms, which work by reducing the amount of free radicals in the body. The modulation of inhibitor kappa B kinase-beta (IKK-) kinase activity about inflammatory processes may give benefits in moderating inflammation, which is an important element in the prevention of cardiovascular disease. IKK facilitates the production of inflammation by phosphorylating IB, increasing the transcription factor nuclear factor-kappa B (NF-B). As a result, earlier research suggests that IKK is a prospective therapeutic target for both the prevention and therapy of cardiovascular disease. Anticoagulant therapy is necessary for a variety of cardiovascular illnesses, including stroke management in atrial fibrillation and the treatment and subsequent prevention of acute coronary syndrome. The phrase hypolipidemic activity, commonly known as lipid-lowering medicine, refers to any chemical that can lower the concentration of lipids and lipoproteins (lipid-protein complexes) in the circulation. Low-density lipoprotein (LDL) and very low-density lipoprotein (VLDL) levels have been associated with an increased vulnerability to cardiovascular diseases such as coronary artery disease, myocardial infarction, and cerebrovascular accident. The examination of the molecular mechanism underlying the possible cardioprotective benefits of herbal medicines using cell and molecular technologies remains unknown. Before beginning any herbal treatment program, it is critical to do a thorough risk assessment to prevent overlooking medication interactions and bad effects connected with such medicines. Furthermore, scientific research into the nano-scale characteristics of phytochemicals is lacking. Natural nanoformulations outperform synthetic or herbal drugs in the prevention and treatment of cardiovascular disease (CVD) (Zuorro, 2021).

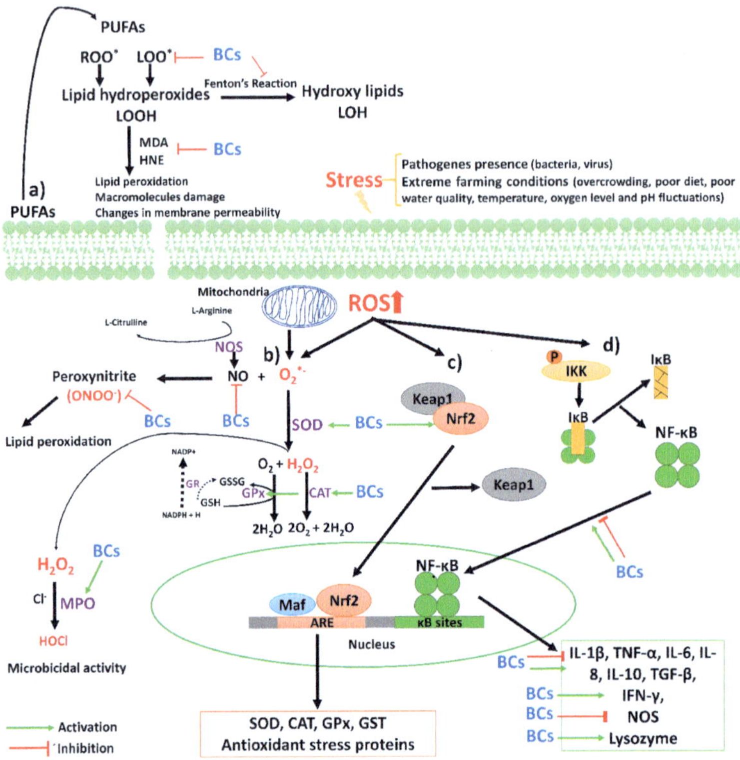

Figure 9.2. Graphical representation of bioactive molecule action.

Source: Leyva-López, N., Lizárraga-Velázquez, C. E., Hernández, C., & Sánchez-Gutiérrez, E. Y. (2020). Exploitation of Agro-Industrial Waste as Potential Source of Bioactive Compounds for Aquaculture. *Foods*, *9*(7), 843. MDPI AG. Retrieved from http://dx.doi.org/10.3390/foods9070843

Several studies have been conducted to explore the effectiveness of natural products and herbal medicines in the prevention or treatment of cardiovascular disease. This drive is influenced by several elements. The concept of a financially feasible treatment plan has been contrasted with the currently established standard of care and the prevalent consensus on its safety and effectiveness. The aforementioned discoveries have led to the use of many botanical species for the treatment of cardiovascular diseases. The heart's protection by evaluating cardiac enzymes such as lactate dehydrogenase (LDH), cardiac protein, lipid peroxidases, and transaminases. According to findings, an extract produced from *Nerium oleander* (NO) has heart-protective effects. Gayathri et al. investigated the possible cardioprotective benefits of NO flowers in animals subjected to isoproterenol-induced oxidative cardiac stress. The study's findings suggested that the plant had a beneficial cardioprotective impact. The current study sought to determine Amaranthus viridis Linn's cardioprotective effects. The study's findings demonstrated a considerable drop in the levels of cardiac enzymes in the treated animals, giving proof for the plant's cardioprotective capabilities. *Ginkgo biloba* seeds and leaves have been shown in studies to have cardioprotective effects. Panda and Naik tested extracts of *Ginkgo biloba* and Ocimum sanctum on mice with isoproterenol-induced cardiac necrosis. Rats with isoproterenol-induced cardiac necrosis were compared to rats with normal blood enzyme levels, and a decrease was found . There are apparent variations in the effects of a combination of 100 mg/kg body weight *Ginkgo biloba* and 50 mg/kg body weight *Ocimum sanctum* and a combination of 50 mg/kg body weight of the former. *Ginkgo biloba* and *Ocimum sanctum* extracts at 100 mg/kg body weight provided considerable protection against cardiovascular disease. *Tinospora cordifolia* extracts obtained from the root, fruit, and stem have been proven to have heart-healthy characteristics. Its cardioprotective activities are related to a range of phytoconstituents (Fernandes da Silva, 2017).

According to empirical evidence, eating habits play a substantial impact in the development of cardiovascular and metabolic diseases. According to study findings, health can be promoted by reducing the consumption of highly processed foods and substituting them with fruits, vegetables, seeds, and legumes. The dietary ingredients are high in unsaturated fats, minerals, fiber, carotenoids, and phenolics, have a low salt level, and do not include any artificial food additives. According to a prior study, coffee and tea are often used for the control of cardiovascular disease. Green tea consumption has been linked to enhanced cardiovascular health in experimental, epidemiological, and clinical research. A previous study has demonstrated this relationship. Nonetheless, the evidence for a difference in the risk of cardiovascular disease between black and green tea is weak. Miura et al. discovered that consuming green tea (1.7 mg catechin/day/mouse) for 14 weeks led to the development of atherosclerosis in mice with hypercholesterolemia and apolipoprotein-deficient animals. The ingestion of green tea in potable water at a concentration of 3.5 g/liter for two weeks appears to have resulted in a drop in blood pressure in hypertensive rodents at risk of stroke. In a second study, diabetic animals with cardiac failure were given green tea at a dose of 300 mg/kg body weight for four weeks to improve antioxidant defense and lipid profile. According to the findings of one particular research study, drinking three cups of tea per day can reduce the risk of acquiring cardiovascular disease. *Camellia sinensis L.* is a plant with a high concentration of phenolic chemicals. The drug in question contains a large level of caffeine. Epicatechin (EC), epigallocatechin (EGC), epicatechin-3-gallate (ECG), and epigallocatechin-3-gallate have been found as phenolic compounds in tea and are known to have therapeutic effects for the treatment of cardiovascular disease.

Drinking coffee has been linked to a lower risk of suicide, neurological illness, and cardiovascular disease. Moderate use of coffee or caffeine has no negative influence on cardiovascular health. Previously, it was theorized that coffee consumption was linked to an increased risk of developing cardiovascular disease. The aforementioned assumption is supported by limited epidemiological research, and coffee consumption may give protection against a variety of cardiovascular diseases. According to research, drinking 35 cups of coffee per day continuously for a year can lead to a 15% reduction in the probability of acquiring cardiovascular disease. Furthermore, no link has been shown between increased coffee consumption and an increased risk of cardiovascular disease (Zuorro, 2021).

Furthermore, drinking 15 cups of coffee every day has been shown to reduce the risk of death when compared to not drinking any coffee at all. Even with continuous use, the risk of cardiovascular disease did not rise in people who had previously experienced a cardiovascular event. Individuals with hypertension and uncontrolled blood pressure, on the other hand, should limit their coffee consumption. Caffeine, diterpenes like cafestol and kahweol, and numerous isomers of chlorogenic acid are all bioactive chemicals found in coffee (Fernandes da Silva2017).

9.4 CANCER PREVENTION

Cancer is a disease characterized by abnormal cell proliferation that has the potential to extend to other body organs and invade them. One of the factors that contribute to the development of cancer is alterations in the genes that control how the body functions. Due to the escalating problem of treatment resistance and the continuously increasing incidence of cancer worldwide, there is a growing interest in various cancer prevention techniques.

Chemoprevention is the use of synthetic, organic, or biological compounds to reduce the risk of cancer in individuals who are otherwise healthy. Chemopreventive drugs either prevent DNA damage, which causes cancer or reverse or impede the division of premalignant cells with DNA damage, thereby inhibiting cancer formation. This method has been demonstrated to be effective in clinical trials for breast, prostate, and colon cancer. Given the continued increase in cancer incidence, the ineffectiveness of standard chemotherapies in controlling cancer, and the excessive toxicity of chemotherapies, it is clear that an alternative method is required. The results of the first trial using tamoxifen to ascertain the efficacy of chemoprevention in breast cancer patients revealed a significant reduction in invasive breast cancer. The effectiveness of chemopreventive medications in preventing cancer in high-risk populations demonstrates the rationality and desirability of the proposed strategy. Several dietary compounds, including capsaicin, cucurbitacin B, isoflavones, catechins, lycopene, benzyl isothiocyanate, phenethyl isothiocyanate, and piperlongumine, have been shown to inhibit the growth of cancer cells, suggesting that they may function as chemopreventive agents (Zuorro, 2021).

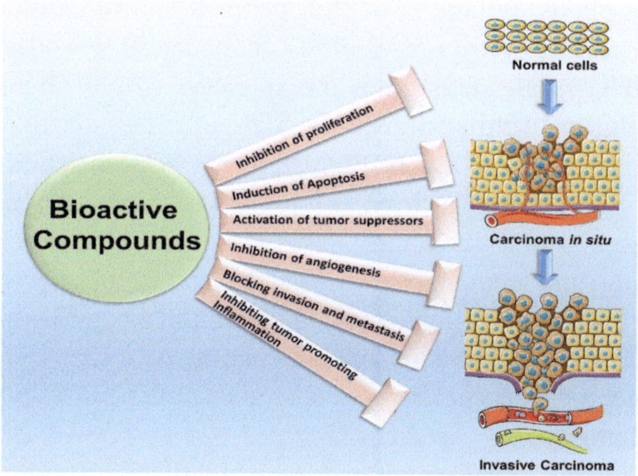

Figure 9.3. Cancer chemoprevention by natural products.

Source: Muniraj N, Siddharth S, Sharma D. Bioactive Compounds: Multi-Targeting Silver Bullets for Preventing and Treating Breast Cancer. *Cancers*. 2019; 11(10):1563. https://doi.org/10.3390/cancers11101563

Chemoprevention is the use of organic, synthetic, or biological agents to halt, block, or reverse the initial phases of carcinogenesis or to prevent the invasion of precancerous cells. With an increased understanding of cancer biology, the identification of molecular targets, and the efficacy of chemoprevention in preventing breast, prostate, and colon cancer, interest in the field of chemoprevention has exploded. On a molecular level, cancer chemoprevention has been distinguished by the modification of multiple pathways that are crucial to the three main phases of carcinogenesis, initiation, promotion, and progression. The Food and Drug Administration or FDA has approved ten innovative medications to treat precancerous lesions and reduce the risk of developing cancer (Fernandes da Silva, 2017).

Clinically, there are three categories of chemoprevention: primary, secondary, and tertiary. Primary chemoprevention is appropriate for both high-risk cancer patients and the general population. Secondary chemoprevention is designed for patients with precancerous lesions that may develop into invasive cancer. Primary chemoprevention has been divided into two distinct categories: primary and secondary. Common chemopreventive agents include nonsteroidal anti-inflammatory medications (NSAID) and phytochemicals. Tertiary chemoprevention, in contrast, attempts to prevent the recurrence of cancer.in breast cancer, tamoxifen therapy is an example of tertiary chemoprevention.

Capsicum plants include both chili peppers and the aromatic alkaloid capsaicin (trans-8-Methyl-N-vanillyl-6-nonenamide). The Scoville Heat Unit (SHU) scale determines the spiciness of chili peppers and the appropriate dilution of chili extracts. The SHU and concentration of capsaicin in a spicy chile pepper are inversely related. The concentration of capsaicin in numerous peppers ranges from 0.1 to 1.0%. Numerous cancer models have demonstrated that capsaicin is chemopreventive, tumor-suppressing, radiosensitizing, and is anticancer. Capsaicin may be an effective treatment for osteoarthritis symptoms when oral nonsteroidal anti-inflammatory drugs are contraindicated due to side effects. Capsaicin interacts with TRPV1 (transient receptor potential cation channel subfamily V member 1), a receptor subfamily. The capsaicin receptor is also known as the TRPV1 receptor. In general, TRPV1 binding does not mediate the anti-cancer effects of capsaicin. Several studies, however, have shown that contact with TRPV1 increases intracellular calcium, resulting in apoptosis. The activator protein 1 (AP-1) and nuclear factor kappa B (NF-B) signaling pathways, as well as the signal transducer and activator of transcription 3 (STAT3) signaling pathways, are activated and, respectively, responsible for tumor growth; capsaicin treatment inhibits these pathways. It has also been demonstrated that capsaicin produces reactive oxygen species (ROS), depolarizes mitochondria, and may cause cell cycle arrest and death. Capsaicin inhibits bladder cancer cell migration by interacting directly with Sirtuin 1 (SIRT1) and decreasing SIRT1 deacetylase. It was discovered that the mortality of pancreatic cancer cells induced by capsaicin was associated with a reduction in -catenin signaling. The administration of 5 mg/kg capsaicin orally to mice with pancreatic tumor implants significantly inhibited their growth. Within one hour of oral administration, the maximal capsaicin concentration and distribution in numerous organs, including the kidneys, lungs, and stomach, are reached (Zuorro, 2021).

Capsaicin has the potential to be used as a cancer therapy because it inhibits carcinogen activity via multiple pathways and elicits apoptosis in a variety of *in vitro* and *in vivo* cancer cell lines. However, animals given natural capsaicin have been observed to develop cancer. According to research, the tumor may have been caused by chemicals that contaminated the capsaicin present in chiles. Capsaicin's irritant properties may have led to an increase in blood flow, which may have increased the absorption of tumor promoters and carcinogens and, consequently, their concentrations, increasing cancer risk. Direct administration of >98% pure capsaicin did not result in any cases of cutaneous malignancy, and all animals were healthy. Numerous small epidemiological studies have found a link between capsaicin consumption and gallbladder or stomach cancer, but their interpretation is complicated by the presence of known carcinogens in foods containing capsaicin. Currently, there is no evidence to support the theory that capsaicin metabolites may cause DNA damage and promote carcinogenesis. Due to its apparent safety and efficacy in animal models, the safety and efficacy of purified capsaicin against cancer can be evaluated in humans (Fernandes da Silva, 2017).

Chemotherapy is typically a cost-effective and risk-free treatment because cancer can be prevented by modifying dietary habits. After the FDA of the United States authorized tamoxifen and raloxifene for reducing the risk of breast cancer, this technique acquired popularity. Numerous epidemiological and preclinical studies have conclusively demonstrated that dietary factors play a role in both the prevention and treatment of cancer. Currently, multiple clinical trials are investigating the chemopreventive effects of the natural compounds listed above. For currently available chemotherapeutic treatments, drug-induced toxicity is a significant obstacle. On the other hand, using natural compounds to prevent cancer may reduce the associated damage. The vast majority of naturally occurring chemopreventive drugs have bioavailability issues. This article discusses natural compounds that act on multiple signaling pathways associated with cancer growth, suggesting that they may be effective anti-cancer drugs.

9.5 DIGESTIVE HEALTH

Due to their high nutrient content and high caloric density, plant-based foods are consumed globally. Their consumption has increased due to a variety of metabolic diseases associated with a diet containing animal products. In addition to their nutritional value, plant foods contain a variety of bioactive compounds with health-promoting properties.

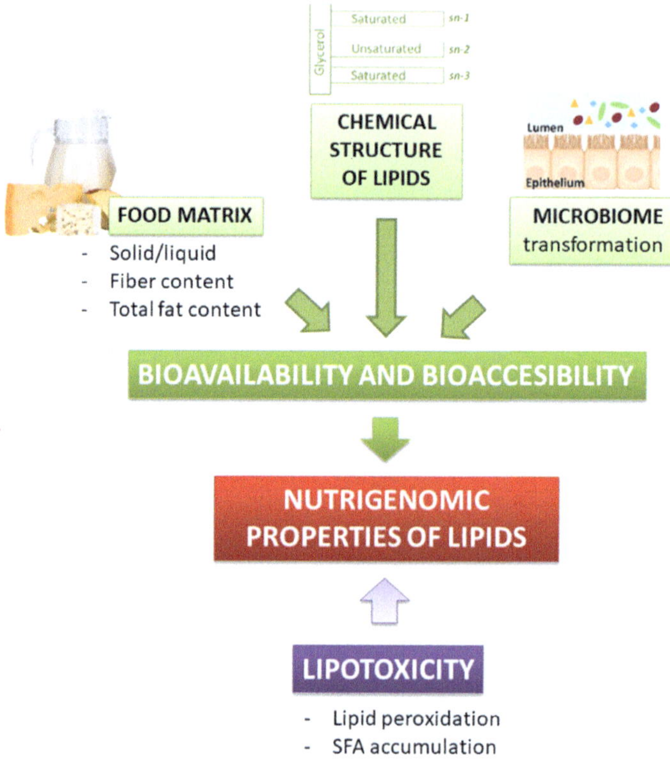

Figure 9.4. Nutrigenomics of lipids.

Source: Bordoni L, Petracci I, Zhao F, Min W, Pierella E, Assmann TS, Martinez JA, Gabbianelli R. Nutrigenomics of Dietary Lipids. *Antioxidants*. 2021; 10(7):994. https://doi.org/10.3390/antiox10070994 [accessed 5 Jun, 2023]

It has been demonstrated that plant-derived bioactive compounds, such as biologically active proteins, polyphenols, phytosterols, biogenic amines, carotenoids, and others, are beneficial to human health in cases of cancer, cardiovascular disease, and diabetes, as well as in individuals with gastrointestinal, immune function, and neurodegenerative disorders. Prior research has demonstrated that bioactive substances possess antioxidative, anti-inflammatory, and immunomodulatory properties, as well as the ability to improve intestinal barrier function, all of which contribute to their ability to mitigate the negative effects of a wide range of human disorders. This study examines the mechanisms underlying the bioactive properties of a variety of these plant components, as well as the bioactive components derived from fruits, vegetables, cereals, and other plant sources with health-promoting properties. This study concentrates predominantly on the potential of bioactive chemicals derived from food, providing researchers with information that may be useful in the development of novel strategies, such as selecting promising bioactive components to construct functional meals for a variety of noncommunicable diseases (Fernandes da Silva, 2017). The intestinal epithelium functions as a formidable barrier to the outside world, preventing the passage of potentially hazardous substances (such as infectious microorganisms) that could provoke an immune response in the host. Consequently, the integrity of this barrier is essential for maintaining appropriate intestinal homeostasis and effective protective responses to chemical and microbial hazards. Increased intestinal permeability increases the number of pathogens, toxins, and allergens that penetrate the body and stimulates inflammation and the immune system. The delicate and precise equilibrium of intestinal homeostasis can be disrupted by several unknown factors governing barrier function. This review will examine the role of dietary supplements (such as probiotics, prebiotics, and synbiotics) in managing intestinal barrier function and regulating gastrointestinal health. We will pay particular attention to three human diseases: food allergy, irritable bowel syndrome, and inflammatory bowel disease (Zuorro, 2021).

The gastrointestinal tract (GIT) has the greatest mucosal surface that is continuously exposed to the outside environment. By operating as a biochemical and physical barrier between the host and the outside environment, a single layer of cells shields the host from potentially irritating and antigenic substances in the luminal compartment. A protein network regulates the epithelial layer that forms this barrier and orchestrates complex biological functions such as the absorption of nutrients, secretion, and transport of water. Connecting epithelial cells, the paracellular pathway permits selective permeability. This permeability is mediated by the complex of junctional proteins. The maintenance of intestinal homeostasis by extension, host health relies heavily on intestinal barriers, which function as a dynamic and responsive interface. The microbiota (previously known as microflora), a vast community of microorganisms that exist in a symbiotic relationship with their host and populate the GIT, is now widely acknowledged as a crucial regulator of intestinal homeostasis and overall gut health with multiple positive effects. A strong intestinal barrier reduces the likelihood of tissue damage and pathogen transmission, thereby retarding the progression of the disease. When the intestinal barrier fails to function properly (leaky gut), pathogenic microorganisms can infiltrate the body. In this context, the factors and conditions that affect intestinal barrier function are of great interest because they may have a significant impact on a variety of intestinal disorders, including inflammatory bowel disease (IBD), irritable bowel syndrome (IBS), and colorectal cancer, as well as other non-gastrointestinal pathologies, including obesity, diabetes, neurological disorders, and food allergies (FA) (Fernandes da Silva, 2017).

The intestine is a complex ecosystem that is exposed to numerous dietary components and commensal microorganisms; interactions in this environment are diverse and frequently synergistic. The production of mucin and other antimicrobial compounds by the epithelium, which limits bacterial colonization and adhesion, influences microbial communities. The food's composition (such as the consumption of prebiotics) can affect the intestinal epithelial barrier and the bacterial populations it supports. In contrast, gut microbiota can play a significant role in intestinal health by protecting the intestinal barrier, modulating the immune system, and stimulating host metabolism. Short-chain fatty acids (SCFAs), essential vitamins, antioxidants, and other active compounds (such as probiotics) released as remnants of microbiota or bacteria during food digestion affect host homeostasis. Dysbiosis is a minor change in the composition of the gastrointestinal microbiota that can lead to several pathophysiological diseases. In addition, it is widely acknowledged that dietary components have beneficial effects on the host that go beyond ordinary nutrition, which has prompted the development of functional food concepts. Several food supplements, for instance, have been developed around the consumption of either live bacteria (i.e., probiotics) or substrates utilized by host microorganisms (i.e., prebiotics), which are the best-characterized bioactive substances in the diet and have been demonstrated to have a positive effect on gut health and overall host well-being (Zuorro, 2021).

9.6 WEIGHT MANAGEMENT

Weight control and obesity prevention are significant priorities for health organizations to avoid metabolic illnessesAccording to the WHO, obesity is a serious public health issue that has a negative impact on social, physical, and psychological health. Obesity is defined by a body mass index (BMI) of 30 kg/m2 and is linked to chronic illnesses, mortality, and morbidity. Childhood obesity is a severe health problem, according to European studies, particularly in Mediterranean nations with high obesity rates. Greece has the highest rates of childhood obesity in Europe. However, because high-energy meals increase energy intake disproportionately, encouraging people to eat fewer of all foods may not have the desired impact. Portion control is essential for weight loss. A better method could be to encourage a balanced diet, such as the Mediterranean diet, as well as more physical activity and the eating of fewer high-energy meals and more low-energy-dense foods.

In the 1980s, the Japanese Ministry of Health and Welfare recognized functional foods (FFs). Functional foods, in addition to their nutritional content, improve specific biological functions, boosting health, happiness, and the avoidance of chronic illnesses. Every day, customers throughout the world consume helpful meals. By 2020, functional meals and beverages are expected to generate $192 billion. Functional foods are often labeled as "healthy" or "natural." Functional meals are those that affect certain biological processes to improve physical health and/or reduce the risk of acquiring chronic illnesses. However, nutritional and health claims about functional foods' capacity to prevent illness are only made when there is sufficient scientific data to back them up. Functional foods may promote health only when incorporated into a well-balanced diet, such as the Mediterranean diet (Fernandes da Silva, 2017).

Food contains trace levels of bioactive chemicals, the effects of which on health are being studied. A high intake of natural functional foods, such as bioactive-rich fruits and vegetables, reduces the risk of chronic diseases such as cardiovascular disease, cancer, metabolic syndrome, type II diabetes, and obesity. A variety of biomolecules, including resveratrol, epigallocatechin gallate, curcumin, oleuropein, sulforaphane, quercetin, ellagic acid, anthocyanins, and b-glucans, have been studied for their potential effects on specific molecular pathways linked to cancer, diabetes, metabolic syndrome, and cardiovascular disease. Because of their various experimental approaches and often disputed results, more clinical and epidemiological research is needed to confirm their efficacy. The major stilbene in the human diet is found in grapes, red wine, and pomegranates. The major polyphenol in green tea, epigallocatechin-3-gallate (EGCG), is an antioxidant and scavenger. Two tri-phenolic groups make up this gallic acid-epigallocatechin ester. Curcuma longa Linnaeus, a ginger family polyphenol, may have anti-inflammatory, antioxidant, and chemopreventive activities. The antioxidant capabilities of the flavonol quercetin (QR), which is found in berries and tea, are provided by three important components: the 4-oxo group, the 2, 3-alkene, 3- and 5-hydroxyl groups, and the B ring hydroxyl groups. Ellagic acid (EA), a polyphenol produced from ellagitannins found in nuts and berries, is a powerful antioxidant and free radical scavenger. Pomegranates, berries, and red grapes contain water-soluble anthocyanins (ACs), which are pigments with purple, red, and blue colors. Anthocyanins may have anti-inflammatory, antioxidant, and chemoprotective properties (Zuorro, 2021).

Obesity is a worldwide epidemic that is connected with several chronic inflammatory or degenerative disorders, as well as metabolic effects. Increased calorie consumption, decreased physical activity, a positive energy balance, and food addiction may all contribute to global obesity. Food addiction is defined as the frequent eating of high-sugar, high-fat meals accompanied by intense desires. Increase your energy expenditure and eat a well-balanced diet to achieve a negative energy balance and control your weight and obesity. Attempts to identify and implement alternative obesity prevention treatments have increased in recent years to reduce metabolic effects and their associated costs. Clinical and epidemiological research stresses the relevance and potential health benefits of functional foods. Combining functional foods with a well-balanced diet can help with weight loss. Gojiberry, cranberry, and pomegranate drinkers had lower BMIs than non-consumers in recent research of 300 healthy volunteers. According to a scientific study, B-glucane, glucomanane, low-fat and sugary meals, and other functional foods and bioactive components may help obese people regulate their weight and metabolic health when taken as part of a balanced diet. However, studies show that when ingested alone and in big quantities, these functional foods and bioactive substances are useless for weight control.

Coffee is the most popular non-alcoholic beverage with pharmacological effects. Green coffee grounds, which are unroasted coffee beans, are one of the most traded grains. Arabic coffee cultivation spread over Latin America, Asia, and Africa centuries ago, beginning in Ethiopia. Both organic and conventional coffee is derived from the *Coffea canephora* and *Coffea arabic*a species. Coffee has become a cultural phenomenon, with hundreds of billions of cups drank each year. The Dietary Guidelines Advisory 2015 recommends two to three cups of coffee per day or 400 mg of coffee in tablet or powder form. In healthy people, moderate coffee use (3-5 cups per day or 400 mg of caffeine per day) was not associated with an increased risk of major chronic illnesses such as cardiovascular disease, premature mortality, or cancer. More than this amount, according to science, is harmful (Fernandes da Silva, 2017).

9.6.1 Coffee bioactive

According to research, those who drink moderate amounts of coffee are less likely to acquire type II diabetes and cardiovascular disease. Caffeine may lessen the risk of Parkinson's disease. Green or roasted coffee may prevent metabolic syndrome (MetS) in hypercholesterolemic and healthy persons by reducing blood pressure, glucose, and triglyceride levels. Another study found that providing 40 male rats with metabolic syndrome, cardiovascular disease, and hepatic structure 68.3 mg/kg coffee for 8-9 weeks enhanced cardiovascular disease. In a randomized, controlled, blinded research, 137 persons who had arrhythmic events were given 500 mg of caffeine daily for two months and had no episodes. However, additional research has shown the potential risks of increasing consumption, including cardiac arrhythmia (Zuorro, 2021).

9.6.2 Weight reduction with coffee/green coffee

Several studies suggest that green coffee may help with weight loss, however, the evidence is mixed. Green coffee bean extract was found to reduce weight and body fat in a recent study. Coffee's antioxidant components may also aid in weight loss and fat reduction. In a randomized controlled experiment, 25 obese men and 95 obese women drank 3 to 5 cups of coffee per day for three weeks and lost weight, and had higher Bifidobacterium spp levels. Separate research employed two coffee varietals, Coffeea Arabica, and Coffea canephora. In various pill formulations, the bioactive components (5-caffeoyl quinic acid) and decaffeinated green coffee extract were both present. Drinks were made by steeping 10-20 m filter sheets in hot water bags for varying periods.

The effects of raw material, extraction duration (5 and 10 min), concentration (2.5, 5, and 10 mg/mL), and coffee type were explored to improve brewing conditions. Coffee drinkers showed decreased weight and body fat. Because of its high chlorogenic acid concentration, green coffee bean extract outperformed roasted coffee and caffeine in lowering body weight as well as epididymal and perineal fat weight in multiple investigations. Coffee may influence the intestinal microbiota of obese individuals and animals. Coffee may be a useful weight-control approach, despite the paucity of data supporting its anti-obesity mechanism. Coffee may help with weight loss. Although further research is needed, several studies have found that coffee and green coffee can help people lose weight, notably by lowering their BMI (Zuorro, 2021).

Functional foods can help with weight loss, prevent obesity, and treat metabolic disorders such as high cholesterol and glucose levels. Numerous epidemiological research, human clinical trials, and animal studies have recently been conducted to investigate the influence of various functional meals and their bioactive constituents on weight control. Many processes have been proposed. Green tea catechins and coffee caffeine may have the ability to regulate BMI and weight. The findings are dubious because earlier research has demonstrated that coffee drinking increases body mass. Coffee and tea's anti-obesity characteristics might be an innovative technique for treating or preventing obesity. Berries and pomegranate boosted fat oxidation, lowered body weight and waist circumference, and protected against the metabolic effects of obesity in human studies. More research is needed to determine the benefits of berry anthocyanins on weight loss. Despite their high-fat content, nuts were related to weight maintenance, weight management, and satiety in the majority of trials. Avocados and olive oil, both abundant in fatty components that may be ingested in moderation, do not affect weight by increasing or decreasing BMI. The results of ginger consumption studies have been inconsistent; some studies have indicated increases in body weight while others have found decreases in BMI (Fernandes da Silva, 2017).

Chapter

10 TECHNOLOGIES FOR EXTRACTION OF PLANT BIOACTIVE COMPOUNDS

CONTENTS

Because bioactive compounds are used in a variety of commercial industries, including the food, pharmaceutical, and chemical sectors, the most effective method for extracting these active compounds from plants is required. Different new approaches, in addition to conventional ones, have been developed to extract bioactive ingredients from plants; however, none of these approaches are currently considered as the standard. The key input parameters are largely responsible for determining the efficiency of conventional and non-conventional extraction methods; comprehending the plant matrix; and bioactive compound chemistry (Falconer, 2017).

10.1. SOXHLET EXTRACTION

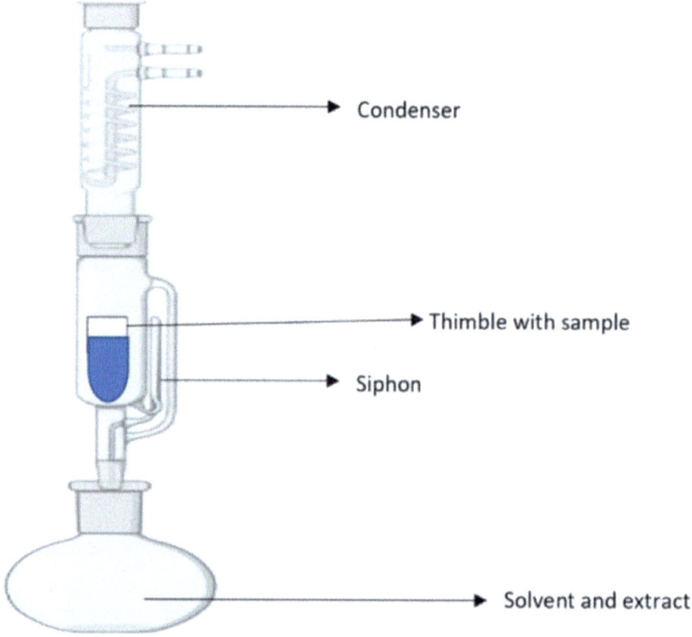

Figure 10.1. Extraction of plant chemicals using the Soxhlet method.

Source: Kumar A, P N, Kumar M, Jose A, Tomer V, Oz E, Proestos C, Zeng M, Elobeid T, K S, et al. Major Phytochemicals: Recent Advances in Health Benefits and Extraction Method. *Molecules*. 2023; 28(2):887. https://doi.org/10.3390/molecules28020887 [accessed 6 Jun, 2023]

Soxhlet extraction is a widely used extraction method for thermally stable analytes. It involves placing a small volume of dry sample on the apparatus through which the solvent flows. The sample is collected in the hot solvent after the extraction solvent is continually cycled along the matrix through boiling and condensation. Different solvents such as dichloromethane (DCM), unadulterated or blended in with CH3)2CO hexane mixtures can be applied. It is not advisable to use nonpolar solvents only. A typical Soxhlet extraction will take less than eight hours (Sit, 2022).

For years, SOX has been the standard analytical procedure, and it is without a doubt the most commonly used method for extracting bioactive compounds from plants. While considered more traditional, SOX is the acknowledged comparison standard and thus of key significance to the US EPA strategy, as mentioned in the EPA Method 3540. First, the sample is set in a disposable thimble which is attached to the Soxhlet device. Refluxing the solvent results in continual extraction. The extract is relocated to the boiling flask once the extraction chamber is full. This procedure is repeated until the extraction process is complete. SOX may be time-consuming whereby an extraction can range from 12 to 24 hours. Furthermore, the majority of extractors make use of a sizable amount of solvents that are filtered out (a few hundred milliliters).

By modifying the Soxhlet process as proposed by Randall during the mid-70s, the extraction period can be significantly reduced. Until the preferred extraction is achieved, the sample is immersed inside the boiling solvent. Since analytes are highly soluble in hot solvents, this speeds up extraction. The thimble is elevated above the boiling solvent following this extraction process until any remaining extract is removed. Integrated sample concentration can be achieved by carefully evaporating the solvent and shutting the solvent return valve. Also, the mechanization of the Randall-Soxhlet technique can be performed more effectively, which prompted the incorporation of the mechanized SOX technique by the U.S. EPA.

The Soxhlet extractor relies on the concept of the "greedy" cup Pythagorean. The liquid inside begins to drain out from a hole at the bottom when the container is filled beyond a certain point. During extraction, solvent vapor flows up the distillery way, into the primary chamber, and up inside the condenser where it consolidates and trickles down. The solid sample will be partially dissolved in the solvent, which will fill the primary chamber. The solvent is returned to the round base flask by the siphon when the chamber is nearly full to restart the process. Every time the extraction process is repeated, a greater amount of the final compound is dissolved, leaving behind the insoluble impurities inside the thimble. Compounds are extracted from the sample in this manner (Falconer, 2017).

A relatively small volume of solvent is re-used to conduct multiple extractions, in contrast to the traditional method. Since significantly less solvent is needed in a Soxhlet extraction as a result, time and money are saved. Additionally, the Soxhlet extractor can operate continually, making it a great choice for extracting components over several hours or days. Franz Ritter von Soxhlet originally developed the instrument to extract lipids from milk solids. Presently, the Soxhlet extractor is applied during exhaustive extractions, especially in the oil and food businesses. Likewise, it is commonly used to extract bioactive compounds from plants, which are important for managing environmental resources. When properly configured, the Soxhlet extractor can run continuously following the below steps:

- Stacking the sample compound comprising the desired component into the thimble
- Adding the thimble into the primary Soxhlet extractor chamber
- Addition of the selected solvent to a round base flask and placement onto a warming mantle
- Connection of the Soxhlet extractor over the round base flask (Sit, 2022)
- Connecting a reflux condenser over the extractor, with cold water passing through the base and leaving above
- Now that the device is established, the next stage comprises warming the dissolvable substance to reflux and extraction at the expected time period.

10.2. MACERATION

It is an extraction method whereby a container is filled with finely grounded plant material, either stem bark, root bark or leaves; Furthermore, the menstruum is poured until it fully covers the drug compound. Among the easiest methods of extraction is maceration, whereby coarse plant material is soaked in solvents like ethanol, ethyl acetate, acetone, methanol, hexane etc. Afterward, the container is sealed and left open for up to three days. This method is a better choice for commercial purposes because of its lower solvent consumption, simplicity, speed, and low cost. Additionally, it improves the yield and extract quality (Falconer, 2017).

The organized tissue is converted into a suspension of whole cells during maceration, which produces pulpy products used as a base for pulp juices and nectars, baby food, and components in dairy products like puddings and yogurts (Pilnik and Voragen, 1993; Khatri et al, 2015). The chemical components of the leaf cells, primarily oxidative, polyphenols, and degradative enzymes, amino acids, lipids, etc., are exposed through maceration, which breaks down the leaf structure and converts these compounds into oxygen.

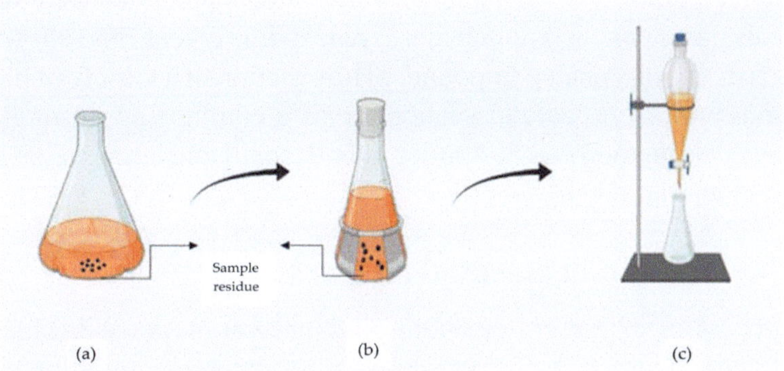

Figure 10.2. Maceration of herbal extracts.

Source: Kumar A, P N, Kumar M, Jose A, Tomer V, Oz E, Proestos C, Zeng M, Elobeid T, K S, et al. Major Phytochemicals: Recent Advances in Health Benefits and Extraction Method. *Molecules*. 2023; 28(2):887. https://doi.org/10.3390/molecules28020887 [accessed 6 Jun, 2023]

More significantly, catechins interact with the enzyme polyphenol oxidase, triggering the process known as "tea oxidation," which has also been wrongly called "fermentation." The conventional rolling, crush, tear, and curl (CTC), as well as the Laurie tea processor (LTP) techniques, are the most prevalent. The strategy used together affects the resultant extract product. In comparison to the two alternative options, conventional rollers result in less destruction of the cell matrix. As a result, conventional maceration results in less oxidative reactions and slower fermentation. The CTC and LTP maceration methods cause more extensive oxidation and greater cell matrix destruction, resulting in less aromatic teas with greater plain tea quality parameters. LTP production requires a milder physical wither compared to CTC processing (Sit, 2022).

Besides, the choosing of proper solvents is significant in the system for the extraction of specific plant extracts. The maceration process comprises the crushing of plant matter into smaller particles to build the surface area for easier blending with solvent and effective extraction of components. The mixture of plant material and solvents is then filtered using a filtration medium after settling for a while, stirred at various intervals, and so on. The effectiveness of the extraction of bioactive compounds from the plant cell relies on the form of solvent and form of plant material. The most significant factor that affects extraction efficiency is the solvents polarity.

For effective extraction, this method makes use of a variety of solvents and time-temperature combinations. Maceration breaks the cell structure and reveals the chemical components to react with solvents and allow for the extraction of various plant compounds. The exaction of a variety of bioactive compounds on a large scale in a laboratory is a common application of this technique. This method can be applied at both small and large scales, finding use at an industrial level.

10.2.1 Maceration in essential oil extraction

In the process of extracting essential oils, maceration is used. Maceration is an extractive method that is performed at room temperature. It comprises immersing the plant in a fluid e.g. oil, alcohol, and water, depending on the plant component and liquid used, for a variable amount of time inside an airtight container. The plant must be thoroughly cleaned and separated from foreign constituents like weeds, topsoil, pebbles or rocks and materials that are not suitable for extraction before processing. Depending on the final product, the plant material may be used fresh or dry (Falconer, 2017).

To increase contact between the plant material being extracted and the fluid (dissolvable), the plant should be sliced into smaller pieces. The solvent won't effectively infiltrate the cells when the pieces are too big. They should also not be ground into a powder; that would result in the loss of the plant's volatile active compounds (essential oils) as well as the complex separation of plant material from the used liquid following maceration.

The chemical nature of the plant's compounds must be taken into consideration when selecting the solvent. When choosing a solvent, solubility and the extraction's intended use should be taken into account. Specifically, considering their solubility and the extraction's intended use. Alcohol is typically the most frequently used substance due to its ability to extract most of the plant's molecules (active compounds), including molecules that are lipophilic, soluble in oil or other organic solvents, or hydrophilic, water-soluble. When only the lipophilic components (fats) need to be separated, vegetable oil is used, however, water is used for the extraction of hydrophilic components only (Sit, N. 2022).

Common plant extracts derived from maceration:

- Infusion: This is the process where a plant, whether fresh or dried, is left to mature in vegetable oil e.g. olive oil or seed oil. Fresh or dried plant samples must be left to macerate in vegetable oil.
- Tinctures: It involves the maceration of plant products in alcohol-water mixture yields
- Glycerin macerates: The liquid comprises a combination of equal amounts of ethyl alcohol, water, and glycerin. The plant parts processed through maceration in this mixture include tissue buds and young sprouts (Falconer, 2017).

10.3. HYDRODISTILLATION

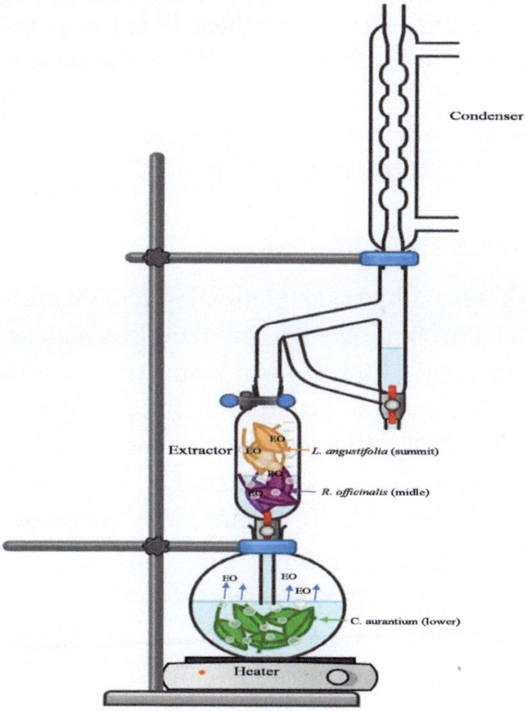

Figure 10.3. Hydrodistillation uses water and steam for extraction.

Source: El Kharraf S, Faleiro ML, Abdellah F, El-Guendouz S, El Hadrami EM, Miguel MG. Simultaneous Hydrodistillation-Steam Distillation of *Rosmarinus officinalis*, *Lavandula angustifolia* and *Citrus aurantium* from Morocco, Major Terpenes: Impact on Biological Activities. *Molecules*. 2021; 26(18):5452. https://doi.org/10.3390/molecules26185452[accessed 6 Jun, 2023]

Another common method for extracting bioactive compounds, mostly essential oils, is hydrodistillation, which uses steam or water. The Clevenger equipment and basic steam distillation are typically used to carry out this method. In hydrodistillation, first, the plant component is pressed into a still chamber, then, water is included in adequate quantities, before being brought to steam.

The extraction of bioactive compounds and essential oils from plant materials has long been achieved through hydro distillation. There are three ways to perform hydrodistillation: water distillation, steam distillation, and direct steam distillation (Azmir et al., 2013). Hydrodiffusion, hydrolysis, and heat decomposition are the three primary physicochemical processes available in hydrodistillation. The primary means by which bioactive components in the plant matrix are extracted and transported are through steam and hot water. Indirect cooling gathers the vapor mixture and separates oil and bioactive mixtures from the water. Dried over anhydrous sodium sulfate, essential oils, and oil-based bioactive compounds are commonly used. Since hydrodistillation is frequently done at temperatures above the water's boiling point, a few volatile components, organic pigmentation, and heat-labile bioactive mixtures may be lost.

10.3.1 Hydrodistillation process

There are three kinds of hydrodistillation: water, steam, and water-steam distillation methods. In hydrodistillation, the plant material is first packed into a still chamber, then sufficient water is added and the mixture is brought to a boil. On the other hand, direct steam is infused into the plant material. The primary factors that influence the release of bioactive compounds from plant tissue are steam and hot water. The oil-water vapor mixture is condensed through indirect water cooling. The condensed mixture flows from the condenser to the separator, where oil and bioactive materials separate consequently from the water (Sit, 2022).

Mother liquors may be produced in large quantities during the hydrodistillation process. Due to its water-soluble phenolics, this liquor adds value in addition to the issue of waste disposal. According to a 2007 study, when used in high concentrations, the ethyl-acetate extract of mother liquors produced through the hydro distillation of oregano may outperform both butyl hydroxyl toluene and ascorbic acid. Furthermore, despite being a good source of antioxidant compounds, the solid filtrate of the hydrodistillation process is typically disposed of. After the hydrodistillation process, the wet residual component—herb and water—may be treated to produce phenolic acid-rich extracts that make full use of the process wastes (Falconer, 2017).

The primary physicochemical processes involved in the hydrodistillation of plant material include: i) Hydrodiffusion, ii) Hydrolysis, and iii) Heat-induced decomposition

Hydrodiffusion

In hydro diffusion, the essential oil is extracted by circulating steam through the plant material. After bringing a few liters of water to a boil in the laboratory, steam rises in a column comprising the roughly ground plant matter. The liquid is gathered in a graduated burette once the vapor phase is channeled to a condenser. The hydrosol moves to the left within the beaker due to a bent connection at the bottom of the burette, whereas the essential oil stays in the burette.

In contrast to the steam distillation method, in which the dry steam cannot penetrate the dry cell membrane, hydro-diffusion involves the passing of hot water and essential oils through dry plant membranes. Since the essential oils are released during the comminution process, plants must be milled before steam distillation. Soaking plant material in water is another way to improve steam distillation because plants' cell membranes are resistant to essential oils, but soaking plant materials in water makes their cell membranes permeable. Additionally, essential oil liquefies in the gland's water when it is boiled. During this procedure, steam is used to transfer the vaporized oil from the oil-water solution entering the cell membranes of plants through osmosis. However, the oil's solubility in water does not affect the rate of essential oil vaporization, nor does the volatility of its components. Milled plant material requires less time to distill than non-milled plant material.

It is possible to measure the volume of extract recovered and measure the yield of plant introduced after around two hours of extraction (measured from the initial condensed drop). This method allows for the recovery of hydrosol. Additionally, it allows for extractions on a greater number of plants. At a much lesser scale, this method is rather similar to the one that is usually employed in industry. With hydro diffusion, however, the extraction may be less comprehensive compared to a Clevenger device.

The inventor of the Clevenger apparatus, Joseph Franklin Clevenger, published in 1928, gave it its name. There are different models such as specific glassware with round-base flask. The jar, of variable sizes, contains boiled water and the extracted plant. The steam falls into the small burette on the right as the assembly's steam rises to a condenser. Oil floats on top of the water, which is slowly returned to the heated flask along the diagonal passage. Following 2 hours of extraction, the extraction volume gathered in the burette can be instantly estimated (Sit, 2022).

In some cases, the Clevenger device provides greater yields compared to hydro diffusion. This could be caused by many factors. Particularly, the packing of plant components in the hydro diffusion column occasionally results in the formation of pockets that are not adequately extracted due to their low steam exposure. Nevertheless, this sort of extraction is somewhat barely applied at a large scale, while the yield is for the most part overestimated and can be generally expected in the extraction plant. Furthermore, although we can adapt the size of round-base flasks, we are substantially more restricted in how much plant material can be separated (Falconer, 2017).

Hydrolysis

Essential oils and water undergo a chemical reaction called hydrolysis. Esters, which are the constituents of essential oils, tend to react with water at high temperatures to create alcohol and acids, however, the reaction is not comprehensive in all directions. When using the water distillation method, adding more water results in more alcohol and acid, which reduces the yield of essential oils. The rate of hydrolysis increases with the length of time that the oil and water are in contact with one another.

Heat decomposition

Every component of essential oils becomes unstable at high temperatures. The hydro-distillation temperature should be low to improve oil quality. Hydro-diffusion, hydrolysis, and decomposition through heat happen simultaneously and influence each other. A common traditional extraction technique is hydro-distillation. This can be used to extract essential oils from plants in diverse ways as follows:

- Solvent extraction

Dissolvable extraction is the most widely used technique, where the material goes through a cycle where the solvent passes through the plant cell wall while the solute disintegrates in the solvents the solute and the extract is finally collected. The solvent that is used to extract medicinal plants is also called menstruum. The type of plant, the plant portion to be extracted, the nature of the bioactive components, and the accessibility of a solvent all affect the solvent of choice. In general, the extraction of polar compounds is done with polar solvents like water, methanol, and ethanol, while the extraction of nonpolar compounds is done with nonpolar solvents like hexane and dichloromethane. During fluid extraction, the traditional way is to choose two miscible solvents like water-dichloromethane, water-hexane, and water-ether. Because of its high polarity and organic solvent miscibility, water is present in all combinations. To facilitate separation, the compound that will be extracted throughliquid–liquid extraction must be soluble in organic solvents but not in water. In addition, the polarity of solvents used in extraction is used to classify them, with n-hexane being the least polar and water is the most polar (Sit, 2022).

During fractionation, the chosen solvent is added according to increased polarity, from the least polar solvent, n-hexane, to the most polar solvent, water. During fractionation, two low-polarity solvents (n-hexane and chloroform) are typically chosen, followed by two medium-polarity solvents (dichloromethane and n-butanol) and another with the greatest polarity (like water).

Solvent extraction properties

a) Water. This is one of the properties of extractive solvents. It is used to extract diverse polar compounds because it is the most polar solvent. Merits: It can dissolve many different substances; it is inexpensive, non-toxic, nonflammable, and extremely polar. Demerits: It can cause microbial and mold growth; it requires a significant ratio of heat to concentrate the extract, and it may result in hydrolysis.

b) Alcohol. It is water-soluble, polar, and capable of extracting polar secondary metabolites. Merits: Alcohol is self-preservative at concentrations above 20%. Because only a small ratio of heat is needed to concentrate the extract, it's nontoxic at low concentrations. Demerits: It cannot dissolve fats, wax, and gums; it is volatile and flammable.

c) Chloroform. It is useful for extracting compounds like flavonoids, fats, terpenoids, and oils because it's a nonpolar solvent. Merits: It has no color, has an appealing smell, and is solvable in alcohol. In addition, the body efficiently metabolizes and absorbs it. Demerits: It is sedative and contains carcinogenic elements.

d) Ether. This is a nonpolar solvent and is helpful in the extraction of mixtures like terpenoids, coumarins, alkaloids, and unsaturated fats. Merits: It is tasteless, water-miscible, and has a low boiling point. Additionally, it is an extremely stable compound that does not react with acids, metals, or bases. Demerits: It is highly flammable and volatile (Falconer, 2017).

e) Ionic liquid (e.g. green solvent). This is a one-of-a-kind extraction solvent that is extremely heat stable and highly polar. Even at 3,000 degrees Celsius, it can remain liquid and be used at high temperatures. It has great water miscibility and is entirely suitable in the extraction of polar mixtures. Merits: It has amazing solvent properties that draw in and produce microwave, and consequently it is reasonable for microwave-aid extraction. It is nonflammable and is valuable for fluid extraction and exceptionally polar. Disadvantage. It is not the best choice for making tinctures.

Popular medicinal plant extraction methods

(i) Digestion. It is an extraction technique for medicinal plants. During the extraction process, this method makes use of a moderate amount of heat. The powdered drug material is then poured into a clean container alongside the extraction solvent. The blend is added over a water shower or in a broiler at temperatures of 50 ^{o} C. Heat is added throughout the extraction process to reduce the consistency of the extraction solvent and improve the release of auxiliary metabolites. Plant materials that are easy to dissolve work well with this method (Sit, 2022).

(ii) Decoctiona. This process involves continual hot extraction using a predetermined water volume as the solvent. A plant material that has been dried, ground, and powdered is put into a clean container. Afterward, water is added and stirred. The extraction is then accelerated by applying heat throughout the process. The procedure usually takes about 15 minutes to complete. Typically, the solvent-to-crude drug ratio is 4:1 or 16:1. It is used to extract plant material that is heat-stable and water-soluble.

10.4. INFUSION

This relies on a similar principle as maceration; both involve immersing the plant matter in boiling or cold water and allowing it to steep. The maceration period for infusion is more limited. A rotary evaporator can also be used to separate and concentrate the liquid in a vacuum. The plant parts to be used are soaked for 15 minutes in boiling water before being filtered using a filter or filter paper to complete the infusion. In the market, there are numerous herbal products stuffed in filter paper appropriate for this process without using extra utensils. The total dosage is about one gram of plant for every ten centiliters of water.

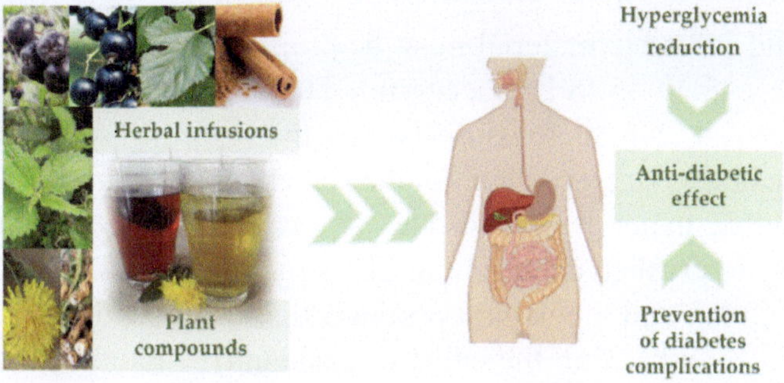

Figure 10.4. Herbal infusions are commonly used for medicinal purposes.

Source: Studzińska-Sroka, E., Galanty, A., Gościniak, A., Wieczorek, M., Kłaput, M., Dudek-Makuch, M., & Cielecka-Piontek, J. (2021). Herbal Infusions as a Valuable Functional Food. Nutrients, 13(11), 4051. MDPI AG. Retrieved from http://dx.doi.org/10.3390/nu13114051

Medical tea is one type of beneficial infusion. It's recommended in psychophysical asthenia, bronchitis, diarrhea, asthma and secondary in the treatment of obesity, arteriosclerosis, and hyperlipidemia. The infusion can be prepared by placing 1 teaspoon of tea leaves or powdered tea in a cup of boiling water and letting it stay for ten minutes (Falconer, 2017).

10.4.1 Background of infusion

Infusion extraction, also known as "washing extraction", "diffusional extraction", "leaching", and "solvent extraction" is the special solution of at least one constituent of a strong blend by contact with a liquid solvent. In this context, it may also refer to the removal of constituents from the insoluble material surface, as well as the discharge of volatiles and aromatics from solvent and insolvent matter through the physical or chemical solvent activity. In addition to the term "physical solution," it may also refer to physical or chemical reaction dissolution.

While the word "solvent" is used throughout this application, it should be understood that the term may refer not only to water but also to ethyl and isopropyl alcohol, steam, petroleum naphtha, acetone, perchlorothylene, hexane, trichloroethylene, chlorinated hydrocarbons, and other solvents applied during the extraction processes. Additionally, the word "solvent" as captured in this application is not limited to solvents only. The general steps in the process of extracting an infusion include reducing the scale of the infusible material's particles, solvent-based infusion extraction of infusible material, and separation of the extract and extraction residue (Sit, N. 2022).

The solid infusible material must be properly prepared before infusion for infusion-extraction to be successful. The solvent must diffuse into the mass of material, similar to osmosis, and the resulting solution or extract must diffuse out, as small particles of the solvent material may be surrounded by an insoluble material matrix. The action will greatly accelerate when these solids are crushed and ground. The structure of animal and vegetable bodies is cellular, and the natural products that can be extracted from them are typically found within the cells. Occasionally, rupturing the cell walls and exposing the material to be dissolved or washed during extraction can be achieved by grinding, crushing, and rolling it to different sizes. Different brewing techniques have been discovered to work best with different coffee "grinds." Nevertheless, grinding accelerates volatility. For example, ground coffee starts to lose its freshness exposed to air due to oxidation of the chemical components that give it flavor and aroma.

As a result, coffee has traditionally been vacuum-packed after grinding. The fragrance and flavor industries use the concentrates and distillates of different natural products i.e. blossoms, leaves, fruits, stems, and roots. Compound flavors and fragrances can only be made with these essential components. According to Teraniski et al., in the "Flavor Research, Principles and Techniques," many of the more significant aromatic components of plants are volatile and exist in rather low concentrations, such as parts per million or billion. This presents a challenge for producers of aromatic extracts. Therefore, it would be advantageous to use an extraction method in which the infusible material is hermetically sealed before and during extraction (Falconer, 2017).

Coffee, for example, swells when it absorbs liquid, slowing the percolation of gravity-fed devices. If pressure is used to force liquid through the ground material, it becomes compressed, subsidizing flow rates even further. When water is permitted or forced to pass across the batch without agitation, preferable routes—also known as "channels" or "tunnels" —may even emerge. There is no uniformity of contact between the solvent and insolvent material in preferred routes or channels, therefore, the soluble constituents are leached from particles adjacent to the channel early on, depleting them early. Later, liquid passing through the channels extracts less, while other particles not adjacent to the channel remain practically un-leached. evices that are solvent through different and infusible materials disregard the notable benefits of the "tumult" process. Therefore, instead of forcing liquid across the infusible material, it could be ideal to mix liquid solvent and insolvent solid material in a receptacle such that the insolvent material is suspended within the solvent before a withdrawal of the liquid extract. Therefore, it would be advantageous to permit the solvent to be moved into the solvent material, agitating and suspending inside the solvent, ensuring more standard contact with the solvent and extraction infusion (Sit, 2022).

It is common knowledge that the infusion extraction process can sometimes be accelerated by relatively higher temperatures. In general, the higher the temperature, the faster, more complete, and higher the extract concentration. The expanded rate is expected, in addition to other factors, to bring down constancy when a fluid solvent is applied. Notwithstanding, it is feasible to get temperatures excessively high for the specific infusion material, sugar beets, for instance, yield unwanted solutes or cause substance decay of the product when handled at too high a temperature. To put it another way, there is an optimal temperature or temperature range for any particular product. For instance, the Pan American Coffee Bureau's non-profit Coffee Brewing Institute, Inc. has suggested that the ideal temperature for brewing coffee is 195° F. It is attractive consequently, to allow close temperature management of the solvent-infusible material blend.

One more significant parameter of the mixture extraction process is the "precarious" time or time span the mixture is mixed with the insolvent material. The infusion time and the period of gravity-induced percolation are critical for other products as well since they are affected by different variables i.e. particle size, container configuration, etc. These may either be too long or short for the best infusion extraction process. Spices, for instance, are made from different plant parts; leaf, fruit, and bark are utilized in their natural state after being dried and ground; spice extracts—most commonly oleoresins—have a significant market. The oleoresins are generally isolated by use of solvents in infusion extractions, and typically by solvent gravity permeation through the ground solid. Different solvents may be used based on the specific spice. The contact period with the solvent is a sensitive factor determining the composition of the concentrate as well as its synthesis, and the dissolvability of certain constituents based on the contact period with the solvent and its temperature. The ratio in which infusion extraction occurs relies on various factors of which limited knowledge is currently available. Solution washing from the external layer of impenetrable solid particles may be quick, requiring just the blending of infusible matter and solvent, though, a solute leaching from the inner section of solids will be generally slowed. The ratio of osmosis or dialysis, dissolution of solutes from the network of insoluble material, and rate of substance response will all vary for various cycles. It is notable, for instance, that the properties of soybean oil vary based on the rather longer or limited filtering periods required. (Treybal, "Mass-Move Activities", McGraw-Slope, 1955, p628). The period required for liquid permeation through a ground material depends on factors such as the specific solid or fluid and the molecular size of the material, yet for a given cluster or solvent, the contact time during permeation is of generally fixed term. It is important in this way, to work with close control throughout the contact hour with infusible material (Falconer, 2017).

The capacity to vary the solvent/infusible material mixture may be beneficial for acquiring useful results, for instance, to help with breaking the infusion material cell walls. Likewise, pressurizing the solvent/batch blend and vacuum extract removal may fast-track the collection of volatiles not otherwise delivered. It is advisable to allow the capacity to apply and control the ideal pressure to help with the imbuement extraction process. In this manner, various factors such as; solvent type, temperature, pressure, steep time, and level of agitation; affect not just the speed and proficiency of the extraction cycle yet also the composition of the particular extract. The capacity to vary each factor or any mix of these factors would be beneficial in various ways, for instance, experimentation on a limited scale to decide the optimal mixture of parameters for extraction of particular infusible material before graduating to financially deliver it (Sit, 2022).

10.5. PERCOLATION

Percolation involves the movement of water through the filtration. Ultimately, as the water percolates inside the inner filtrate layers, it reaches the ground-level water, which is water underneath the surface. The upper surface of this underground water is known as the "water table". Extraction happens when a liquid (solvent) moves through a plant. This is an extractive process that is done at room temperature and that in a real sense means "passing the liquid through a solid matter drop by drop." Percolation is more effective compared to maceration since it is a continuous cycle, where the saturated solvent is continually replaced by new solvents. Zhang et al. contrasted the saturation and refluxing extraction strategies to extract *Undaria pinnatifida*. They discovered that the components in the primary part, fucoxanthin, from the permeation extraction technique was higher compared to the refluxing strategy while there was no huge contrast in extract production between the two strategies (Falconer, 2017).

A percolator is the equipment used in this procedure. It is a glass container with openings at both sides and the shape of a narrow cone. A dried, crushed, and finely ground plant material is saturated with the extraction solvent in a clean compartment. The mixture is preserved for four hours before more solvent is added. Afterward, the contents are put into a percolator with the bottom closed and left to stand for 24 hours. Next, the extraction solvent is transferred from the top until the drug substance is completely soaked. The lower some portion of the percolator is opened, and the fluid is permitted to dribble gradually. Continually, a small amount of solvent was added, and the solvent was pushed through the drug material and down during the extraction by gravitational force. When the volume of added solvents reached 75% of the proposed volume for the entire preparation, the solvent addition was stopped. After filtration and decantation, the extract is separated. Thereafter, the marc is expressed, and the final ratio of solvent is added to reach the needed volume.

Goupi patch is a component Chinese medication heavily uses comprising of 29 traditional Chinese medications. Fu et al. used the complete alkaloid content not defined by acid-base titration as the index, streamlining the ethanol permeation technique as splashing the medication with 55% alcohol for a day and afterward percolating up to 12x the quantity of 55% alcohol. While utilizing the extricating ratio of sinomenine and ephedrine hydrochloride as the primary index, Gao created another optimized percolation strategy: soaking the medication with 70% ethanol for a day and percolating with up to 20x the ratio of 70% ethanol. The transfer ratio of sinomenine and ephedrine hydrochloride respectively measured at 78.23 and 76.92% (Sit, 2022).

10.5.1 Percolation concept

Percolation theory provides a definition of gelation that doesn't reject the development of closed circles and thus doesn't forecast a different concentration for large groups. Since it considers the establishment of closed loops, the measured permeation threshold values are higher compared to those derived from classical theory and are closer in comparison to those observed through experimentation. Every site or cross-sectional point on a square grid can be regarded as a utilitarian group that can bond between adjoining units. A random bond filling that ties different cross sections can be expected. In case p addresses the small number of bonds in the filled lattice, as one builds p, an -ever-increasing number of units connect. Whenever p is higher compared to the critical value P_c, an infinite lattice-spanning group appears (Falconer, 2017).

The solvent is gradually passed across the plant, loading itself up with active compounds, and is gradually driven away by additional unadulterated solvent that is increased from above. Before putting the plant into the percolator, it should be well shredded but be mindful so as not to make the particles excessively fine. This would cause complexity in isolating it from the extraction fluid (solvent), making the concentrate cloudy and contain buildup at the base. Nevertheless, it is encouraged to preventively saturate or dampen the plant material with the preferred solvent for final extraction. This allows for the expansion of plant cells and promotes the passing of active compounds across the fluid.

Afterward, the well-prepared plant is placed inside the percolator. The solvent is added from above and moves through the plant compound at a speed defined by the plant going through extraction. The flow speed should not be inflated, otherwise, the solvent won't have the opportunity to penetrate the plant cells and to extract the molecules found inside the plants; similarly, if the flow is slower, more solvents would be required to finish extraction. In general, the flow speed for around 1kg of plant material should not be more than 5 ml/minute.

The solvent choice varies, in view of the chemical context of the active compounds to be extracted. In general, a hydro-alcoholic solvent is applied, and that implies the solvent is comprised of a blend of water and alcohol. This results in more effective extraction processes since the water keeps the plant walls hydrated, whereas the alcohol is chemically comparable in context to most active compounds. Moreover, the existence of alcohol is additionally useful conserving the acquired extraction is concerned. The derived liquid, rich in plant active compounds, is known as leachate. When the percolation is complete, the plant material for extraction, still absorbed by the solvent, goes through a squeezing to recover the residual fluid absorbed within the plant material, this is afterward included in the leachate (Sit, 2022).

Percolation is an extractive strategy that isn't suggested for plants that expand a lot, for instance, plants containing pectin or mucilage, which hinder the solvent from freely flowing. Percolation ends after the colorless liquid leaves the percolator and, becomes devoid of the plant's active compounds. The first leachate collected is described as the head extraction and comprises the highest quantity of extracted content. This is used for progressive partial leachates that will have reduced quantities of active compounds (Falconer, 2017).

10.6 NEW BOTANICAL EXTRACTION TECHNOLOGIES

a) Drying Technology

Occasionally, herbal extracts, particularly aqueous concentrates, must be dried into a solid structure for more extended shelf life. The restructured and controlled drying process can deliver top-quality products with the least loss of active compounds and flavors while preserving microbial safety. Common dry technologies used in the food sector include, for example, vacuum belt drying, drum drying and spray drying, among others. All factors constant, more developed drying technologies such as microwave-vacuum drying, freeze-drying and infrared drying are covered in this section.

b) Infrared Drying

Infrared (IR) radiation using frequencies that vary from 2.5 to 200 μm is a new technique applied to different thermal processing activities in the food business. The application of IR radiation for food dehydration has many benefits. Other than requiring less time, the drying system can be accurately managed whereas the temperature is spread consistently in the drying structure, which highly improves the finished product's quality. It has likewise been found that thermal processing with infrared can expand the number of antioxidants and phenolic mixtures of natural extracts like grape, peanut shell, and citrus, when applied in solvent extraction procedure, i.e. microwave and ultra-sonication (Falconer, 2017).

c) Freeze Drying

Freeze drying, otherwise called lyophilization, is a drying procedure where the liquid medium (typically water) in samples is solidified at low temperatures and afterward transferred from the solid condition into a gaseous state. Unlike traditional drying technologies, the critical benefits of freeze drying include the maintenance of morphological structures and the chemical and biological elements and a high recovery of unstable and heat-sensitive compounds. Regardless of the great benefits, freeze-drying is among the most costly drying processes due to its high-energy use and significant expenses on both operations and maintenance. Currently, freeze-drying is applied in an extensive range of daily food products such as dairy products, instant coffee, and fruit & vegetables (Sit, 2022).

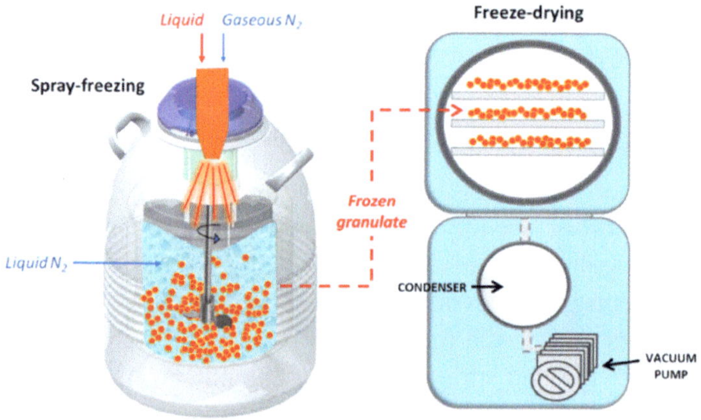

Figure 10.5. Freeze drying of organic plant material.

Source: Bhatta, S., Stevanovic Janezic, T., & Ratti, C. (2020). Freeze-Drying of Plant-Based Foods. Foods, 9(1), 87. MDPI AG. Retrieved from http://dx.doi.org/10.3390/foods9010087

d) Microwave vacuum drying

Microwave vacuuming is a dehydration process combining microwave radiation and vacuum processes. The microwave radiation energy can warm the samples quickly by changing the electromagnetic energy to thermal power. Notwithstanding, the vacuum decreases the water's boiling point, keeping the product temperatures low, and developing a pressure gradient that promotes the drying rate. In contrast with freeze-drying, it is more practical, as the drying process is a lot quicker and consequently permits for higher throughput for similar plant features.

10.7 FRACTIONATION AND SEPARATION

Additional fractionation and separation are occasionally required to deliver products consisting of target compounds with greater purity. Frequently chosen based on the various chemical and physical properties of different ingredients, some common separation technologies include; chromatographic procedures, centrifuge, membrane filtration, distillation, and liquid-liquid extraction. Supercritical liquid extraction is additionally used for separation because it is effective in the extraction of solid raw products, but also to perform fractionation for liquid concentrates. The below sections primarily focus on chromatography, liquid-liquid extraction and membrane filtration (Falconer, 2017).

Column Chromatography

The chromatographic methodology is selective, varied, and the most broadly used strategy in extract fractionation. The separation model of chromatography is based on particle size, charge, solvency, and polarity. In the food business, different macroporous polymeric saps are the most well-known chromatography products used for increased processing, particularly in juice creation, aging, and dairy. Additionally, resin chromatography has been effectively used for demineralization, decolorization, debittering, as well as eliminating poisons and pesticides and improving color, polyphenols, starches, and proteins and with superior calibers. While this process has high selectivity in comparison to other separation techniques, the restricted scale-up range of chromatography is an inconvenience. To fill this deficiency, well-engineered processes executing chromatographic separation, for example, Simulated Moving Bed (SMB) have been created in past few years, which further increase chromatography production and minimize costs.

Membrane Filtration

Being a processing and separation technique, membrane filtration is a form of green technology gaining massive popularity in the food sector. The pressure-run membrane processes consist of, ultrafiltration, microfiltration and nano-filtration including reverse osmosis as per the membrane's pore size. With various pore sizes (or sub-molecular weight cut-offs) of membranes, separation of compounds with a massive molecule scale is possible. Nevertheless, the membrane separation rule isn't just about the pore sizes, yet also the molecular charge of particles and their affinity towards the filtration membrane. This often makes membrane separation complex but also gives greater flexibility of application. Membranes have a few benefits over traditional technologies. For instance, the scale-up structures are highly mechanized and -intuitive, and permit room-temperature operations. Smaller filtrations can likewise be used to eliminate additional solvents without heat treatment as in a traditional evaporation process (Sit, 2022).

Continuous Countercurrent Column

Liquid-liquid extraction, otherwise called solvent extraction, is a renowned separation process that isolates components by using an inconsistent circulation of the parts between two immiscible liquid stages. Mix setters, continuous countercurrent columns and centrifugal contactors are significant scale-up activities in this technology. In continuous extraction column processes, the scattered drops move counter-current against the second phase flow, then the solute-rich stream is gathered either from the top or bottom based on whether the extract solution is lighter or heavier compared to the carrier solvent (Falconer, 2017).

10.8 EXTRACTION METHODS

Although new technologies have improved extraction efficiency and quality, reduced costs, and utilize more environmentally-friendly methods, conventional solvent extraction is still extensively used in the food sector. The differences in the mixture's physical or chemical properties, such as molecular size, polarity, and electrical charges, serve as the basis for extraction methods for developing botanical extracts.

Freeze-thaw-assisted extraction

The freeze-thaw method is considered a sustainable method for the breakdown of cell membranes and tissue in plants. The freeze-thaw pretreatment can significantly increase the level of permeabilization of cell membranes, which makes the subsequent extraction more effective. It has been demonstrated that this technology has the potential to increase yield.

Pressurized Hot Water Extraction (PHWE)

Although traditional water extraction frequently lacks efficiency and selectivity, it is among the cleanest and most environmentally-friendly technologies. An extraction method known as PHWE makes use of superheated water as the extraction solvent. This water is heated to temperatures higher than the atmospheric boiling point of 0.1 MPa, and lower than its supercritical value of 22.1 MPa. Oftentimes, PHWE has been used to obtain relatively polar extracts (Falconer, 2017).

REFERENCES

1. Assunção, J., Guedes, A. C., & Malcata, F. X. (2017). Biotechnological and pharmacological applications of biotoxins and other bioactive molecules from dinoflagellates. *Marine Drugs*, *15*(12), 393. Azmir, J., Zaidul, I. S. M., Rahman, M. M., Sharif, K. M., Mohamed, A., Sahena, F., ... & Omar, A. K. M. (2013). Techniques for extraction of bioactive compounds from plant materials: A review. *Journal of food engineering*, *117*(4), 426-436.
2. Bernhoft, A. J. A. B. (2010). A brief review on bioactive compounds in plants. *Bioactive compounds in plants-benefits and risks for man and animals*, *50*, 11-17.
3. Bernhoft, A., Siem, H., Bjertness, E., Meltzer, M., Flaten, T., & Holmsen, E. (2010). Bioactive compounds in plants–benefits and risks for man and animals. *The Norwegian Academy of Science and Letters, Oslo*.
4. Bizuneh, G. K. (2021). The chemical diversity and biological activities of phytoalexins. *Advances in Traditional Medicine*, *21*(1), 31-43.
5. Bubalo, M. C., Vidović, S., Redovniković, I. R., & Jokić, S. (2018). New perspective in extraction of plant biologically active compounds by green solvents. *Food and Bioproducts Processing*, *109*, 52-73.
6. Carpena, M., Nuñez-Estevez, B., Soria-Lopez, A., & Simal-Gandara, J. (2020). Bee venom: an updating review of its bioactive molecules and its health applications. *Nutrients*, *12*(11), 3360.

7. Cavazos, A., & Gonzalez de Mejia, E. (2013). Identification of bioactive peptides from cereal storage proteins and their potential role in prevention of chronic diseases. *Comprehensive Reviews in Food Science and Food Safety*, *12*(4), 364-380.
8. Chiurchiù, V., Leuti, A., & Maccarrone, M. (2018). Bioactive lipids and chronic inflammation: managing the fire within. *Frontiers in immunology*, *9*, 38.
9. da Silva, R. F., Carneiro, C. N., de Sousa, C. B. D. C., Gomez, F. J., Espino, M., Boiteux, J., ... & Dias, F. D. S. (2022). Sustainable extraction bioactive compounds procedures in medicinal plants based on the principles of green analytical chemistry: A review. *Microchemical Journal*, 107184.
10. Gfeller, D., Michielin, O., & Zoete, V. (2013). Shaping the interaction landscape of bioactive molecules. *Bioinformatics*, *29*(23), 3073-3079.
11. Ghosh, A. K., & Brindisi, M. (2016). Achmatowicz reaction and its application in the syntheses of bioactive molecules. *RSC advances*, *6*(112), 111564-111598.
12. Gutte, K. B., Sahoo, A. K., & Ranveer, R. C. (2015). Bioactive components of flaxseed and its health benefits. *International Journal of Pharmaceutical Sciences Review and Research*, *31*(1), 42-51.
13. Hum, J., & Boccaccini, A. R. (2012). Bioactive glasses as carriers for bioactive molecules and therapeutic drugs: a review. *Journal of Materials Science: Materials in Medicine*, *23*, 2317-2333.
14. Jha, A. K., & Sit, N. (2022). Extraction of bioactive compounds from plant materials using combination of various novel methods: A review. *Trends in Food Science & Technology*, *119*, 579-591.
15. Joana Gil-Chávez, G., Villa, J. A., Fernando Ayala-Zavala, J., Basilio Heredia, J., Sepulveda, D., Yahia, E. M., & González-Aguilar, G. A. (2013). Technologies for extraction and production of bioactive compounds to be used as nutraceuticals and food ingredients: an overview. *Comprehensive Reviews in Food Science and Food Safety*, *12*(1), 5-23.
16. Kendall, A. C., & Nicolaou, A. (2013). Bioactive lipid mediators in skin inflammation and immunity. *Progress in lipid research*, *52*(1), 141-164.
17. Khaw, K. Y., Parat, M. O., Shaw, P. N., & Falconer, J. R. (2017). Solvent supercritical fluid technologies to extract bioactive compounds

from natural sources: A review. *Molecules*, *22*(7), 1186.

18. Kowalczewski, P. Ł., Olejnik, A., Świtek, S., Bzducha-Wróbel, A., Kubiak, P., Kujawska, M., & Lewandowicz, G. (2022). Bioactive compounds of potato (Solanum tuberosum L.) juice: From industry waste to food and medical applications. *Critical Reviews in Plant Sciences*, *41*(1), 52-89.
19. Kowalska, I., Mołdoch, J., Pawelec, S., Podolska, G., von Cossel, M., Derycke, V., ... & Oleszek, W. (2022). Environmental and cultivar variability in composition, content and biological activity of phenolic acids and alkylresorcinols of winter wheat grains from a multi-site field trial across Europe. *Journal of Cereal Science*, *107*, 103527.
20. Kumar, A., Agarwal, D. K., Kumar, S., Reddy, Y. M., Chintagunta, A. D., Saritha, K. V., ... & Kumar, S. J. (2019). Nutraceuticals derived from seed storage proteins: implications for health wellness. *Biocatalysis and Agricultural Biotechnology*, *17*, 710-719.
21. Liu, R. H. (2013). Dietary bioactive compounds and their health implications. *Journal of food science*, *78*(s1), A18-A25.
22. Luiz de Oliveira da Rosa, W., Machado da Silva, T., Fernando Demarco, F., Piva, E., & Fernandes da Silva, A. (2017). Could the application of bioactive molecules improve vital pulp therapy success? A systematic review. *Journal of Biomedical Materials Research Part A*, *105*(3), 941-956.
23. Martins, T. P., Rouger, C., Glasser, N. R., Freitas, S., de Fraissinette, N. B., Balskus, E. P., ... & Leão, P. N. (2019). Chemistry, bioactivity and biosynthesis of cyanobacterial alkylresorcinols. *Natural Product Reports*, *36*(10), 1437-1461.
24. Martirosyan, D., & Ekblad, M. (2022). Functional foods classification system: exemplifying through analysis of bioactive compounds. *Functional Food Science*, *2*(4), 94-123.
25. Maruyama, N., Mikami, B., & Utsumi, S. (2011). The development of transgenic crops to improve human health by advanced utilization of seed storage proteins. *Bioscience, biotechnology, and biochemistry*, *75*(5), 823-828.
26. Miller, S. R., Heurtaux, D., Baati, T., Horcajada, P., Grenèche, J. M., & Serre, C. (2010). Biodegradable therapeutic MOFs for the delivery of bioactive molecules. *Chemical communications*, *46*(25), 4526-4528.
27. Moseti, D., Regassa, A., & Kim, W. K. (2016). Molecular

regulation of adipogenesis and potential anti-adipogenic bioactive molecules. *International Journal of molecular sciences*, *17*(1), 124.

28. Musso, G., Cassader, M., Paschetta, E., & Gambino, R. (2018). Bioactive lipid species and metabolic pathways in progression and resolution of nonalcoholic steatohepatitis. *Gastroenterology*, *155*(2), 282-302.
29. Nicolaou, A., Pilkington, S. M., & Rhodes, L. E. (2011). Ultraviolet-radiation induced skin inflammation: dissecting the role of bioactive lipids. *Chemistry and physics of lipids*, *164*(6), 535-543.
30. Nowak, W., & Jeziorek, M. (2023, January). The Role of Flaxseed in Improving Human Health. In *Healthcare* (Vol. 11, No. 3, p. 395). MDPI.
31. Nwachukwu, I. D., Luciano, F. B., & Udenigwe, C. C. (2013). The inducible soybean glyceollin phytoalexins with multifunctional health-promoting properties. Food research international, 54(1), 1208-1216.
32. Orona-Tamayo, D., Valverde, M. E., & Paredes-López, O. (2019). Bioactive peptides from selected latin american food crops–A nutraceutical and molecular approach. *Critical reviews in food science and nutrition*, *59*(12), 1949-1975.
33. Parikh, M., Netticadan, T., & Pierce, G. N. (2018). Flaxseed: Its bioactive components and their cardiovascular benefits. *American Journal of Physiology-Heart and Circulatory Physiology*.
34. Pedras, M. S. C., Yaya, E. E., & Glawischnig, E. (2011). The phytoalexins from cultivated and wild crucifers: chemistry and biology. *Natural product reports*, *28*(8), 1381-1405.
35. Qin, D., & Xi, J. (2021). Flash extraction: An ultra-rapid technique for acquiring bioactive compounds from plant materials. *Trends in Food Science & Technology*, *112*, 581-591.
36. Rani, A., Saini, K. C., Bast, F., Mehariya, S., Bhatia, S. K., Lavecchia, R., & Zuorro, A. (2021). Microorganisms: A potential source of bioactive molecules for antioxidant applications. *Molecules*, *26*(4), 1142.
37. Ross, A. B. (2012). Present status and perspectives on the use of alkylresorcinols as biomarkers of wholegrain wheat and rye intake. *Journal of nutrition and metabolism*, *2012*.
38. Sangiorgio, P., Errico, S., Verardi, A., Moliterni, S., Tamasi, G., Rossi, C., & Balducchi, R. (2023). Bioactive Lignans from Flaxseed: Biological

Properties and Patented Recovery Technologies. *Nutraceuticals*, *3*(1), 58-74.

39. Shim, Y. Y., Gui, B., Arnison, P. G., Wang, Y., & Reaney, M. J. (2014). Flaxseed (*Linum usitatissimum* L.) bioactive compounds and peptide nomenclature: A review. *Trends in food science & technology*, *38*(1), 5-20.
40. Sobolev, V. S., Khan, S. I., Tabanca, N., Wedge, D. E., Manly, S. P., Cutler, S. J., ... & Gloer, J. B. (2011). Biological activity of peanut (*Arachis hypogaea*) phytoalexins and selected natural and synthetic stilbenoids. *Journal of agricultural and food chemistry*, *59*(5), 1673-1682.
41. Stanger, L., & Holinstat, M. (2023). Bioactive lipid regulation of platelet function, hemostasis, and thrombosis. *Pharmacology & Therapeutics*, 108420.
42. Sun, X., Sun, Y., Li, Y., Wu, Q., & Wang, L. (2021). Identification and Characterization of the Seed Storage Proteins and Related Genes of *Cannabis sativa* L. *Frontiers in Nutrition*, *8*, 678421.
43. Tian, S., Sun, Y., Chen, Z., & Zhao, R. (2020). Bioavailability and bioactivity of alkylresorcinols from different cereal products. *Journal of Food Quality*, *2020*, 1-6.
44. Udenigwe, C. C., & Aluko, R. E. (2012). Chemistry and bioactivity of food phytoalexins. *Bioactive molecules in plant foods*, 13-130.
45. Usman, I., Hussain, M., Imran, A., Afzaal, M., Saeed, F., Javed, M., ... & A. Saewan, S. (2022). Traditional and innovative approaches for the extraction of bioactive compounds. *International Journal of Food Properties*, *25*(1), 1215-1233.
46. Wang, H., Wang, J., Guo, X., Brennan, C. S., Li, T., Fu, X., ... & Liu, R. H. (2016). Effect of germination on lignan biosynthesis, and antioxidant and antiproliferative activities in flaxseed (*Linum usitatissimum* L.). *Food Chemistry*, *205*, 170-177.
47. Zhang, Y., Tao, L., Li, S., & Wei, Y. (2011). Synthesis of multiresponsive and dynamic chitosan-based hydrogels for controlled release of bioactive molecules. *Biomacromolecules*, *12*(8), 2894-2901.
48. Ziegler, S., Pries, V., Hedberg, C., & Waldmann, H. (2013). Target identification for small bioactive molecules: finding the needle in the haystack. *Angewandte Chemie International Edition*, *52*(10), 2744-2792.

INDEX